Fire Code Essentials

Based on the 2018 International Fire Code®

International Code Council
Kevin H. Scott

FIRE CODE ESSENTIALS:
Based on the 2018 International Fire Code®

Kevin H. Scott

International Code Council

Executive Vice President and Director of Business Development:
  Mark A. Johnson

Senior Vice President, Business and Product Development:
  Hamid Naderi

Vice President and Technical Director, Products and Services:
  Doug Thornburg

Senior Marketing Specialist:
  Dianna Hallmark

ISBN: 978-1-60983-786-0

Cover Design: Lisa Triska
Project Editor: Greg Dickson
Project Head: Steve Van Note
Publications Manager: Mary Lou Luif
Publications Production: Sue Brockman

COPYRIGHT © 2018 International Code Council, Inc
ALL RIGHTS RESERVED.

This publication is a copyrighted work owned by the International Code Council, Inc. ("ICC"). Without advance written permission from ICC, no part of this publication may be reproduced, distributed or transmitted in any form or by any means, including, without limitation, electronic, optical or mechanical means (by way of example, and not limitation, photocopying or recording by or in an information storage retrieval system). For information on use rights and permissions, please contact: ICC Publications, 4051 Flossmoor Road, Country Club Hills, IL 60478. Phone 1-888-ICC-SAFE (422-7233).

The information contained in this document is believed to be accurate; however, it is being provided for informational purposes only and is intended for use only as a guide. Publication of this document by the ICC should not be construed as the ICC engaging in or rendering engineering, legal or other professional services. Use of the information contained in this book should not be considered by the user to be a substitute for the advice of a registered professional engineer, attorney or other professional. If such advice is required, it should be sought through the services of a registered professional engineer, licensed attorney or other professional.

Trademarks: "International Code Council," the "International Code Council" logo, "ICC," the "ICC" logo, "International Fire Code," "IFC" and other names and trademarks appearing in this book are registered trademarks of ICC, and may not be used without permission.

Errata on various ICC publications may be available at www.iccsafe.org/errata.

First Printing: May 2018

PRINTED IN THE U.S.A.

Contents

Preface ix
About the International Code Council xi
About the International Fire Code xi
Acknowledgments xii
About the Author xiii
Prerequisite Reading—Occupancy Classification xiv

PART I: CODE ADMINISTRATION AND ENFORCEMENT 1

Chapter 1: Introduction to Building and Fire Codes 2
Code Development 3
The Building and Fire Codes—Scope 5
  International Building Code (IBC) 6
  International Residential Code (IRC) 7
  International Wildland-Urban Interface Code (IWUIC) 8
  International Mechanical Code (IMC) 8
  International Fuel Gas Code (IFGC) 9
  International Property Maintenance Code (IPMC) 9
International Fire Code 9
  Applicability of the IFC 11
  Retroactive construction requirements in the IFC 13
  Change of use or occupancy 14
  Historic buildings 15
  Referenced codes and standards 16

Chapter 2: Legal Aspects, Permits and Inspections 18
Code Adoption 19
  Adoption of the IFC 19
  Amending the IFC 19
  Appendices 21
  Local and state laws 21
Authority 22
  Authority and duties of the fire code official 22
  Technical assistance 24
  Alternative materials and methods 24
  Authority at fires and other emergencies 26
Permits 27
  Operational and construction permits 27
  Construction documents 29
  Permit application 29
  Fees 33
Inspections 33
  Right of entry 33
  Liability 35
  Testing and operation 37
  Unsafe buildings 37
  Stop work order 38
Board of Appeals 39

PART II: GENERAL SAFETY REQUIREMENTS — 41

Chapter 3: General Precautions against Fire — 42
Combustible Materials — 43
- Combustible waste — 43
- Outdoor pallet storage — 44

Ignition Sources — 45
Open Flames — 45
Vacant Premises — 46
Indoor Displays — 48
Hazards to Fire Fighters — 48
Rooftop Gardens and Landscaped Roofs — 48
Mobile Food Preparation Vehicles — 49

Chapter 4: Emergency Planning and Preparedness — 50
Emergency Forces Notification — 51
Public Assemblies and Events — 51
Crowd Managers — 51
Fire Safety and Evacuation Plans — 52
Emergency Evacuation Drills — 54
Employee Training and Response — 55

PART III: SITE AND BUILDING SERVICES — 57

Chapter 5: Fire Service Features — 58
Fire Apparatus Access Roads — 59
- Appendix D — 60

Access to Buildings — 62
Hazards to Fire Fighters during Access — 63
Fire Protection Water Supplies — 64
- Appendix B — 65
- Inspection and maintenance — 65

Emergency Responder Radio Coverage — 66

Chapter 6: Building Systems — 70
Fuel-Fired Appliances — 71
Mechanical Refrigeration — 75
Elevators — 76
Commercial Cooking Operations — 79
- Commercial kitchen exhaust hoods — 79
- Cooking oil storage — 81
- Automatic fire-extinguishing systems — 82

Emergency and Standby Power Systems — 83
Emergency Lighting — 87
Solar Photovoltaic Power Systems — 88
Stationary Fuel Cell Power Systems — 89

Energy Storage Systems 90
    Stationary storage battery systems 90
    Capacitor energy storage systems..........................92

Chapter 7: Interior Finish and Decorative Materials 93
Purpose of the Requirements 94
Interior Wall and Ceiling Finish and Trim 96
Foam Plastics 99
Upholstered Furniture and Mattresses 99

PART IV : FIRE/LIFE SAFETY SYSTEMS AND FEATURES   101

Chapter 8: Requirements for All Fire Protection Systems.... 102
Where Are Fire Protection Systems Required? 103
Construction Documents and Acceptance Testing 106
Inspection, Testing and Maintenance 108
Fire Protection System Impairment 110
Fire Protection System Monitoring 113

Chapter 9: Automatic Sprinkler Systems..................... 114
Level of Exit Discharge and Fire Area 115
Design and Installation Standards 117
Types of Automatic Sprinkler Systems 120
Occupancies Requiring Automatic Sprinkler Protection ... 122
Fire Sprinklers "Throughout" and Exempt Locations 129
Fire Department Connection 130

Chapter 10: Fire Alarm and Detection Systems............... 132
Design and Installation Standards 133
Fundamental Components 136
Occupancies Requiring Fire Alarm and Detection Systems . 139
Carbon Monoxide Alarms 143
Gas Detection Systems 144

Chapter 11: Means of Egress 145
Introduction to Means of Egress 146
Occupant Load 149
Egress Width 152
Exit Access and Exit Access Travel Distance 153
Exit Signs and Means of Egress Illumination 158
Means of Egress Maintenance 160

PART V: SPECIAL PROCESSES AND BUILDING USES — 163

Chapter 12: Motor Fuel-Dispensing Facilities and Repair Garages — 164
- Applicable Requirements by Fuel — 165
- Dispensing Operations and Devices—All Fuels — 165
- Flammable and Combustible Liquid Fuel Dispensing — 168
- Liquefied Petroleum Gas Dispensing — 174
- Hydrogen Dispensing — 176
- On-Demand Mobile Fueling — 178

Chapter 13: Flammable Finishes — 179
- Types of Flammable Finishing Processes — 180
- Spray Booth and Spray Room Construction — 182
- Mechanical Ventilation — 185
- Illumination — 187
- Interlocks — 188
- Fire Protection — 188

Chapter 14: High-Piled Combustible Storage — 190
- What Is High-Piled Combustible Storage? — 191
- Commodity Classification — 192
- High-Piled Combustible Storage Areas — 195
- Storage Methods — 200
- Aisles — 204

Chapter 15: Other Special Uses and Processes — 206
- Combustible Dust-Producing Operations — 207
- Fire Safety during Construction and Demolition — 210
- Welding and Other Hot Work — 212
- Higher Education Laboratories — 215
- Marijuana Grow and Processing Facilities — 218

PART VI: HAZARDOUS MATERIALS — 221

Chapter 16: General Requirements For Hazardous Materials — 222
- Material Classification — 224
- Hazardous Materials Reporting — 229
- Storage and Use — 231
- Maximum Allowable Quantity Per Control Area — 233
- Control Areas — 239
- Hazard Identification Signs — 241
- Separation of Incompatible Materials — 243

Chapter 17: Compressed Gases 246
Cylinders, Containers and Tanks 248
Pressure Relief Devices 249
Markings 251
Security 253
Valve Protection 254
Separation from Hazardous Conditions 254
Exhausted Enclosures and Gas Cabinets 256
Leaks, Damage or Corrosion 256
LP-Gas Cylinder Exchange Program 257
Liquid Carbon Dioxide for Beverage Dispensing 259

Chapter 18: Flammable and Combustible Liquids 261
Classification of Liquids 262
Containers, Portable Tanks and Stationary Tanks ... 264
Design and Construction of Storage Tanks 267
Storage Tank Openings 272
Normal vent 272
Emergency vent 273

Glossary 277
Index 281

Preface

Fire code enforcement is an important public safety function because unwanted fires injure and kill thousands annually. Unwanted fires also inflict a monetary impact on communities because fires remove businesses from the tax rolls while the damaged building is rebuilt. Statistics confirm that over 40 percent of the businesses that experience a fire never reopen because they lose their customer base. Of concern to any community is the accidental release of hazardous materials because of the potential for fire, explosion or injury due to incapacitation by the chemical's constituents. All of these incidents require a response by the fire department, which places fire fighters in danger, especially when an interior rescue and fire attack is required. Given the broad scope of hazards in society, the job of enforcing the fire code is challenging. This is especially true when dealing with hazardous materials, high-piled combustible storage and combustible dust-producing operations.

Fire Code Essentials: Based on the 2018 International Fire Code® (IFC®) was developed to address the need for an illustrated text explaining the basics of the fire code. It is intended to provide an understanding of the proper application of the code to the most commonly encountered hazards found in many communities and cities. The text is presented and organized in a user-friendly manner with an emphasis on technical accuracy and clear noncode language. The content is directed to fire service professionals, code officials, designers and others in the building construction industry.

The content of *Fire Code Essentials* is organized to correspond with the arrangement of the 2018 IFC. It commences with a review of the legal aspects associated with the adoption and enforcement of the fire code provisions, including permitting, right of entry, and inspector liability. It progresses through common hazards that can be found in any occupancy; site and building features that must be addressed with any new construction; fire and life safety systems; hazards presented by special processes and operations; and it concludes with a review of the most commonly encountered hazardous materials. This format is useful to readers because it pulls together related information from the various sections of the IFC into one convenient location while providing a familiar frame of reference to those with code enforcement experience. The book is formatted to follow the steps of new building construction or renovation as well as areas of focus during any fire inspection. This format and arrangement offers the reader a means to understanding why fire code enforcement is an important public safety function and why it is so important to the safety of emergency responders.

Anyone involved in the design, construction or inspection of buildings or industrial processes and hazards will benefit from this book. Beginning and experienced fire inspectors, plans examiners, contractors, engineers, architects, environmental specialists, health and safety professionals, and students in fire science, fire protection, and building inspection technology curriculum or related fields of study and work will gain a fundamental understanding and practical application of the frequently used provisions of the 2018 edition of the IFC.

Reasonable and correct application of the code provisions is enhanced by a basic understanding of the fire code development process, the scope, intent, and correlation of the family of the International Codes, and the proper administration of those codes. This fundamental information is provided in the opening chapters of this book. The book also explains the interaction of the fire code with other local and state regulations. Because the content is focused on the fire code, the book includes prerequisite reading that is important in understanding the *International Building Code* occupancy classification system, how buildings are assigned occupancy classifications, and how these classifications are used in the application of the IFC.

This book does not intend to cover all provisions of the IFC or all of the accepted materials and methods for the construction of fire protection systems, fire and life safety features, or the storage and handling of combustible and hazardous materials. Focusing in some detail on the most common hazards that are found in nearly every community affords an opportunity to fully understand the basics without exploring every variable and alternative. This is not to say that information not covered is any less important or valid. This book is best used as a companion to the IFC and appropriate National Fire Protection Association standards, which should be referenced for additional information.

Fire Code Essentials features full color illustrations and photographs to assist the reader in visualizing the application of the code requirements. Practical examples, simplified tables and highlights of particularly useful information also aid in understanding the provisions and determining code compliance. References to the applicable 2018 IFC sections are cited to assist readers in locating the corresponding code language and related topics in the code.

ABOUT THE INTERNATIONAL CODE COUNCIL

The International Code Council is a member-focused association. It is dedicated to developing model codes and standards used in the design, build and compliance process to construct safe, sustainable, affordable and resilient structures. Most U.S. communities and many global markets choose the International Codes. ICC Evaluation Service® (ICC-ES®) is the industry leader in performing technical evaluations for code compliance fostering safe and sustainable design and construction.

Headquarters:
500 New Jersey Avenue, NW, 6th Floor,
Washington, DC 20001-2070

Regional Offices:
Birmingham, AL; Chicago, IL; Los Angeles, CA
1-888-422-7233
www.iccsafe.org

ABOUT THE INTERNATIONAL FIRE CODE

The *International Fire Code®* (IFC®) is a comprehensive, stand-alone model code that regulates minimum fire safety requirements for new and existing buildings, facilities, storage and processes. The IFC addresses fire prevention, fire protection, life safety and the storage and use of hazardous materials in new and existing buildings, facilities and processes. The IFC provides a total approach of mitigating hazards in all buildings and sites, regardless of the hazard being located indoors or outdoors.

The IFC contains criteria for design, construction and maintenance. For example, before one constructs a building, the site must be provided with an adequate water supply for fire-fighting operations and a means of building access for emergency responders in the event of a medical emergency, fire or natural or technological disaster. Depending on the building's occupancy and uses, the IFC regulates the various hazards that may be housed within the building, including refrigeration systems, application of flammable finishes, fueling of motor vehicles, storage of high-piled combustible materials and storage and use of hazardous materials. The IFC sets forth minimum requirements for these and other hazards and contains requirements for maintaining the life safety of building occupants, the protection of emergency responders, and to limit the damage to a building and its contents as the result of a fire, explosion or unauthorized hazardous material discharge.

ACKNOWLEDGMENTS

Scott Stookey, previously a Senior Technical Staff member with the International Code Council and currently with the Austin, Texas, Fire Department, authored the 2009 *Building Code Basics: Fire*. That book was the basis for this document series. The 2012, 2015 and 2018 editions have built on that original document. *Fire Code Essentials* is the result of a collaborative effort, and the author is grateful for the valuable contributions by the following talented staff of ICC: Hamid Naderi, P.E., C.B.O., Senior Vice President Product Development; Doug Thornburg, A.I.A., C.B.O., Vice President and Technical Director of Product Development; Senior Technical Staff Stephen Van Note, C.B.O and Technical Staff/Education Terrell Stripling for their contributions throughout this project. As always, their assistance and guidance on various provisions were extremely helpful.

ABOUT THE AUTHOR

Kevin H. Scott
President
KH Scott & Associates LLC

Kevin Scott is President of KH Scott & Associates LLC. Kevin has extensive experience in the development of fire safety, building safety and hazardous materials regulations. Kevin has actively worked for over 30 years in the development of fire code, building code and fire safety regulations at the local, state, national and international levels. Kevin previously worked as a Senior Regional Manager with the International Code Council, and before that, he was Deputy Chief for the Kern County Fire Department, California, where he worked for 30 years. He has developed and presented many seminars on a variety of technical subjects including means of egress, high-piled combustible storage, hazardous materials, and plan review and inspection practices.

Kevin was a member of the original IFC Drafting Committee that worked to create the first edition of the IFC. He served for seven years on the IFC Code Development Committee and was chairperson for the committee from 2001 to 2004. Kevin has actively participated in numerous technical committees to evaluate specific hazards and technologies, and to create regulations specific to those hazards. Some of the more significant committees follow:
- High-piled Combustible Storage Committee
- Hydrogen Gas Ad Hoc Committee
- Task Group 400
- Technical Advisory Committee on Retail Storage of Group 'A' Plastic Commodities
- Underwriters Laboratories Fire Council.

Kevin's constant work to improve fire and life safety has been recognized on many levels. His contributions have been acknowledged by various organizations with the presentation of the following awards:
- Mary Eriksen-Rattan Award in 2013—presented by the Southern California State Fire Prevention Officers' Association
- William Goss Award in 2009—presented by the California State Firefighters Association
- Fire Official of the Year Award in 2005—presented by the California Building Officials
- Robert W. Gain Award in 2003—presented by the International Fire Code Institute.

PREREQUISITE READING—OCCUPANCY CLASSIFICATION

Before readers of this book proceed into its content, they must understand that most communities regulate their buildings based on the occupancy classification, which is assigned based on the use and character of a building. A building's use is evaluated for life safety and fire risks, and its character represents the functions and activities that are expected to occur in the building. An occupancy classification is based on the relative hazards within a building, and similar uses are grouped into occupancy categories. A correct occupancy classification establishes the foundation for all the code requirements that are intended for the building's safe use.

Occupancies are classified into groups and subgroups using the requirements in the *International Building Code*® (IBC®). In most communities, the fire code official does not have the legal authority to assign an occupancy classification; this task is normally assigned to the building code official. The IBC addresses not only fire and life safety aspects, but also includes requirements for accessibility of mobility-impaired persons; building sanitation such as potable and wastewater systems; building ventilation such as the fresh air supply and heating, ventilating and air conditioning systems as well as various structural loads of the building itself and external loads including snow, wind, rain and seismic ground movements. A building's occupancy classification influences these and other building code provisions. The *International Fire Code*® (IFC®) is primarily concerned with control of combustible materials and ignition sources; proper design, construction and maintenance of fire protection systems; safety of emergency responders and mitigation of processes or uses that represent a fire hazard or a high potential of injury or death, such as the release of hazardous materials, through safe design, construction, operation and maintenance.

The factors that govern the classification of a building's use must be carefully considered so that those uses or occupancies having approximately the same combustible content and similar fire and life hazard characteristics will be classified under the same occupancy heading. Occupancies should be grouped so that fire protection requirements and height and area limitations applicable to the occupancy groups are rational for all building uses within that group. Every classification must be based on the premise that the uses covered by each will have similar fire hazards and life safety problems and that they share like characteristics. Within any given occupancy group or subgroup, no wide differentiation should exist between the fire hazards of the most hazardous and the least hazardous uses.

The occupancy groups include 10 major classifications as follows:

A Assembly
B Business
E Educational
F Factory-Industrial
H Hazardous
I Institutional
M Mercantile
R Residential
S Storage
U Utility and Miscellaneous

In addition to these major classifications, the occupancy groups of Assembly, Factory-Industrial, Hazardous, Institutional, Residential and Storage are further divided into subgroups in order to accommodate some variations in the hazards associated with the uses within each group (for example, hotel versus an apartment dwelling in the Residential classification). The fire load characteristics in Factory-Industrial and Storage occupancies vary considerably depending on the product or process involved and, therefore, these uses are further classified into subgroups of low and moderate hazard, depending on the potential fire severity.

The occupancy subgroups for specific classifications are as follows:

A Assembly
- A-1 Fixed seating for entertainment, i.e., theater or concert hall
- A-2 Drinking and dining establishments
- A-3 General assembly classification if others don't apply
- A-4 Indoor sports facility
- A-5 Outdoor sports facility

F Factory-Industrial
- F-1 Moderate hazard factory – manufacture or assembly of combustible products
- F-2 Low hazard factory – manufacture or assembly of noncombustible products

H Hazardous
- H-1 Use or storage of hazardous materials with a detonation potential
- H-2 Use or storage of hazardous materials with a deflagration potential
- H-3 Use or storage of hazardous materials which present a significant physical hazard
- H-4 Use or storage of hazardous materials which present a health hazard

H-5 Semiconductor fabrication facilities or research labs

I Institutional
- I-1 24-hour care where a supervised environment or custodial care is provided
- I-2 24-hour medical care or hospital facility
- I-3 Detention facility or jail
- I-4 Day care facility

R Residential
- R-1 Hotel or motel – transient stay
- R-2 Apartment or dormitory – nontransient stay
- R-3 General residential classification if other classifications do not apply
- R-4 Halfway house or group home

S Storage
- S-1 Moderate-hazard storage – combustible products
- S-2 Low-hazard storage – noncombustible products

As more and more buildings are being designed either for a single specialized purpose or as a part of a larger type of building complex, the need for more special code considerations has been recognized. Some examples of these special uses include automobile parking structures, domed stadiums, high-rise buildings, covered mall and open mall buildings, airport terminals and large industrial complexes such as steel mills and assembly plants. For additional information or details of the various occupancy classifications, refer to Chapters 3 and 4 of the *International Building Code*.

PART 1

Code Administration and Enforcement

Chapter 1: Introduction to Building and Fire Codes

Chapter 2: Legal Aspects, Permits and Inspections

CHAPTER 1
Introduction to Building and Fire Codes

Building and fire codes are a group of regulations that address the construction, alteration, maintenance and use of buildings. The *International Fire Code*® (IFC®) is unique because it is the only code that establishes requirements for the property on which a building is located. The IFC requirements address buildings, including access roadways and water supplies for fire-fighting operations. The fire code contains certain limitations on how an individual uses his or her property, as it grants the fire code official the authority to prohibit locations where certain hazardous materials may be stored and to limit the height and location of combustible materials stored outside. It authorizes the code official to regulate site development and construction materials for buildings located in an area that can be impacted by a wildland fire.

Commonly called construction codes, the separate volumes include requirements for the design of the structure itself based on the materials used to erect the building and the internal and external forces it may face, such as a seismic ground movement, wind load, flooding, and heavy roof loading because of snow or rain. Construction codes have extensive requirements to limit or control the development and spread of an unwanted fire and to ensure the safety of the occupants and the ability to exit the building. A major focus of these codes is the health and safety of the building occupants, including requirements for delivering drinking water and removing human and food waste for proper treatment and disposal, as well as heating, cooling, and electrical systems. Building construction codes also establish requirements for protecting the building envelope from weather and from fires that may initiate in, or spread to, a building.

The *International Fire Code* is one of several construction codes published by the International Code Council® (ICC®). A major theme of the IFC is the protection of people. The IFC is concerned with the employees or visitors within the building, the occupants in buildings or uses adjacent to the building, and the emergency responders who must enter the building during an emergency. The fire code requirements are designed to protect the public from the various hazards that are present in a variety of building uses and activities, as well as the safe storage and handling of hazardous materials. The IFC establishes the minimum requirements for fire department apparatus access to buildings, fire fighter access into buildings, fire-fighting water supplies and provisions that address fire hazards inside and outside of buildings. The IFC references the companion International Codes® (I-Codes®) for the construction of buildings regulated by the *International Building Code*® (IBC®), and one- and two-family dwellings and townhouses, which are regulated by the *International Residential Code*® (IRC®). It also references a variety of national standards, such as those from the National Fire Protection Association (NFPA) or Underwriters Laboratories (UL), which provide specific installation or product details that might not be included in the codes.

This chapter reviews how the IFC and the other codes published by ICC are developed and the scope of the other companion codes, and concludes with an extensive review of the IFC scope.

CODE DEVELOPMENT

Construction technology, methods, materials, equipment and processes are constantly evolving. Accordingly, the I-Codes are systematically revised at periodic intervals to keep up with technical changes and to address the improved understanding of hazards. The I-Codes are revised and updated through an open process that invites participation by all stakeholders and affected parties. This code

> **You Should Know**
>
> **cdpACCESS™**
> - cdpACCESS is ICC's cloud-based system for the code development process (cdp).
> - It was developed to increase participation in the code development process.
> - cdpACCESS allows users to create, collaborate, review, submit and vote (if eligible) on code change proposals and public comments.
> - After the Committee Action Hearings, ICC members can view and vote on motions for those code changes that received an assembly motion.
> - After the Public Comment Hearings, ICC posts the Online Governmental Consensus Vote. The proposals and hearing testimony are available to be viewed by everyone; ICC Governmental Member Voting Representatives and Honorary Members are able to vote. •

development process can involve exhaustive research, review, discussion and debate of the potential changes.

A code change begins with the submittal of a proposal to add, modify, or delete a code provision. Any interested individual or group (other than ICC staff) may submit a code change proposal and participate in the proceedings in which it, and all other proposals, are considered. Proposed code changes are submitted online through the ICC cdpACCESS™ website (code development process access). Following the publication and distribution of the proposals, an open public hearing, the Committee Action Hearing, is held before a committee comprised of persons from across the country representing the construction industry and government, including code officials, contractors, builders, architects, engineers, industry professionals, and others with expertise related to the applicable code or portion of the code under consideration. This open debate and broad participation before the committee reflects a consensus of the construction community and those impacted by building and fire codes in the decision-making process. The committee may approve, modify or disapprove the code change proposal.

After the committee vote, the attendees have the opportunity to make a motion from the floor that is different from the motion approved by the committee. If this assembly floor motion is successful, then the assembly floor motion becomes available for online balloting after the Committee Action Hearing concludes. All members of ICC can vote in the Online Assembly Floor-Motion Ballot. If an assembly floor motion receives a majority vote, then the motion becomes a Public Comment to be heard at the Public Comment Hearing. At this point in the process, the first round of the code change cycle is completed and the results of the Committee Action Hearing and the subsequent online voting are compiled and made available as the Report of Committee Action Hearing (ROCAH). Following the availability of the ROCAH, anyone may submit a public comment proposing to modify or overturn the hearing results.

The next public hearing is the Public Comment Hearing, in which the merits of code change proposals that received public comment are debated along with any successful floor motions. Though any interested party may offer testimony, only ICC Governmental Member Voting Representatives (designated public safety officials of a government jurisdiction responsible for administering and enforcing codes) and ICC Honorary Members are permitted to cast votes at the Public Comment Hearing. Public safety officials have no vested financial interest in the outcome, and they legitimately represent the public interest. Votes are cast electronically at the Public Comment Hearing, and the results are then made available online at cdpACCESS. The Public Comment Hearing is followed by the Online Governmental Consensus Vote. Again, in the Online Governmental Consensus Vote only the ICC Governmental Member

Voting Representatives and ICC Honorary Members are permitted to vote. The voting at the Public Comment Hearing and the Online Governmental Consensus Vote are compiled and determine the revisions to the next edition of the code.

A new edition of the code is published every three years. Each I-Code is developed during a single twelve-month cycle during the three-year period. There are two separate code change cycles within each three-year period, with individual codes assigned to one of the two cycles. This important process ensures that the International Codes reflect the latest technical advances and address the concerns of those throughout the industry in a fair and equitable manner (Figure 1-1).

FIGURE 1-1 2018 International Codes

THE BUILDING AND FIRE CODES—SCOPE

Each of the International Codes has common features that are consistent across the entire library of construction and property maintenance codes. Each code begins by stating its scope. The scope establishes the range of buildings, facilities, construction, equipment, and systems to which the particular code applies. The scope of the IFC is different from the other construction codes because its provisions apply to buildings, storage, processes, and hazards located indoors or outdoors of new and existing buildings. The other construction codes are generally limited to new buildings or new

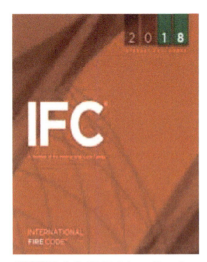

FIGURE 1-2 2018 *International Fire Code*

renovations and additions to existing buildings; when the other construction codes regulate uses or systems located outdoors, the requirements are limited to a particular building service such as plumbing or fuel gas systems.

The International Codes contain cross references to the other applicable codes based on the system or element of a building. For example, the IFC (Figure 1-2) makes extensive references to the requirements in the *International Mechanical Code®* (IMC®) for refrigeration systems, hazardous exhaust mechanical ventilation systems, spray booth ventilation, and commercial cooking operations. The IFC defers to the IBC construction requirements for new buildings or buildings undergoing renovation. Fire code officials generally are not well-versed in the structural design of buildings or the design of the mechanical, plumbing, and fuel gas systems, so the IFC defers to the IBC, IMC, *International Plumbing Code®* (IPC®) or *International Fuel Gas Code®* (IFGC®) in these instances. The IBC charges the building official as the individual responsible for these and other elements regulated by the IBC, including allowable height and area, fire-resistance-rated construction, structural integrity, and accessibility. Any new or existing buildings that are altered or changed must comply with the requirements in the IFC and the IBC. **[Ref. 102.4]**

All of the I-Codes refer to nationally recognized standards. Standards establish the minimum requirements or criteria for the acceptance of materials, components, assembled equipment or machinery, and systems. Each standard referenced by the IFC is considered part of the IFC and must be complied with. If an instance arises where a conflict exists between the code and a standard, the provisions in the IFC supersede the requirements in the standard. The IFC requirements take precedence regardless of whether the requirement is more restrictive or not. Many times, the code requirement is intentionally different from the standard. This occurs because the voting membership decided to modify a provision in the standard—to make it more or less restrictive perhaps, or to apply it to specific situations. The precedence is based on the hierarchy of the code and standard adoption process. **[Ref. 102.7.1]**

Each of the International Codes contains one or more appendices. Appendices are not mandatory requirements unless they are specifically adopted by the governmental entity (see Chapter 2).

International Building Code (IBC)

The provisions of the *International Building Code®* (IBC®) apply to the construction, alteration, maintenance, use, and occupancy of all buildings and structures except detached one- and two-family dwellings and townhouses and their accessory structures, which are covered by the *International Residential Code*. In addition to structural components and systems, the IBC classifies the occupancy of the building and provides requirements for safe interior finishes applied to walls, ceilings, and floors; means of egress; accessibility for mobility impaired individuals; passive and active fire protection; weather resistance, and interior environments (Figure 1-3). These requirements are based on the use and occupancy of the building. The occupancy classification of a building or use is based on its intended function. The IBC prescribes various requirements based on the intended uses within the building and controls the design accordingly. IBC provisions limit the building's height and area, location on property, means of egress, construction and degree of fire protection, and the provisions vary greatly among covered malls, high-rise buildings, warehouses, nightclubs, schools, apartments and grocery stores.

> **You Should Know**
>
> The IFC applies to new and existing buildings and conditions that are hazardous to life, property or public welfare inside or outside a building. Its requirements are correlated to other International Codes. •

FIGURE 1-3 IBC-regulated commercial building

International Residential Code (IRC)

The requirements in the *International Residential Code®* (IRC®) apply to the construction, alteration, use, and occupancy of detached one- and two-family dwellings and townhouses. Such buildings are limited to not more than three stories above grade in height, and each dwelling unit must have a separate means of egress (Figure 1-4). This construction code includes provisions for structural elements, fire and life safety, indoor air quality, energy conservation and the building's mechanical, electrical and plumbing systems. The IRC requires compliance with prescriptive construction provisions or the use of performance design criteria. The provisions in the IFC are applicable to the exterior elements of IRC-regulated buildings, including premises identification, fire apparatus access and water supplies. The IFC administrative, operational and maintenance provisions are also applicable to buildings regulated by the IRC. [Ref. 102.5]

FIGURE 1-4 The IRC sets forth requirements for construction and alteration of single-family dwellings.

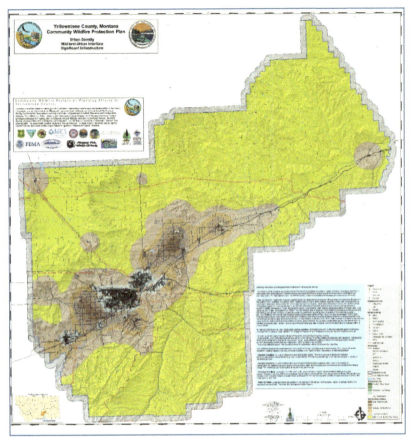

FIGURE 1-5 Wildland-urban interface map (*Courtesy of Billings [MT] Fire Department*)

International Wildland-Urban Interface Code (IWUIC)

The *International Wildland-Urban Interface Code*® (IWUIC®) sets forth requirements for geographical areas where structures and other human development meet or intermingle with wildland or vegetative fuels (Figure 1-5). IWUIC requirements are applied to mitigate the risk to life and property loss from wildland fire exposures and fire exposures from adjacent structures, and to limit the potential of a structure fire igniting adjacent wildland areas. The IWUIC accomplishes this by requiring the jurisdiction to legally identify and map its wildland-urban interface areas followed by applying risk-based construction requirements to improve the ignition resistance of buildings and structures. The IWUIC also prescribes requirements for establishing and maintaining defensible space around buildings.

International Mechanical Code (IMC)

FIGURE 1-6 The cooking appliance and exhaust hood in this commercial kitchen are regulated by the IMC and IFC.

The *International Mechanical Code*® (IMC®) regulates the design, installation, maintenance, and alteration of building mechanical systems that are used to control the environment and related processes. The IMC does not apply to the installation of fuel gas distribution piping and appliances; these systems are regulated by the provisions in the *International Fuel Gas Code*.

The IFC makes extensive reference to the requirements in the IMC for refrigeration systems, commercial kitchen cooking systems, fuel oil piping, hazardous exhaust ventilation systems used in battery rooms, spray booths, semiconductor fabrication facilities, and buildings storing and handling hazardous materials (Figure 1-6).

International Fuel Gas Code (IFGC)

The *International Fuel Gas Code®* (IFGC®) regulates the installation of natural gas and liquefied petroleum gas (LP-gas) systems, fuel gas utilization equipment (appliances), gaseous hydrogen systems, and related accessories (Figure 1-7). The scope of the IFGC extends from the utility company's point of delivery to the appliance shut-off valve. Its requirements address pipe sizing and arrangement, approved materials, installation, testing, inspection, operation and maintenance. The equipment installation requirements include combustion and ventilation air, approved venting and connections to the fuel gas system. IFC Chapter 61 references the IFGC requirements for LP-gas.

FIGURE 1-7 Natural gas meters and the piping on the discharge side of each meter are regulated by the IFGC.

International Property Maintenance Code (IPMC)

The *International Property Maintenance Code®* (IPMC®) establishes minimum regulations for the maintenance of property. Its purpose is to adequately protect the public safety, health and general welfare of individuals who may be affected by the continued occupancy of buildings and the maintenance of structures and premises. Existing structures and premises that do not comply with these provisions must be altered or repaired to provide a minimum level of health and safety. The provisions of the IPMC apply to all existing residential and nonresidential structures. It includes minimum requirements for safe and sanitary living and work conditions including maintenance of lighting, ventilation, heating, sanitation and protection from the elements, fire and life safety (Figure 1-8).

FIGURE 1-8 A leaking sewage system would require repair under the provisions in the IPMC.

INTERNATIONAL FIRE CODE

The provisions of the IFC address the hazards of fire and explosion arising from the storage, handling, or use of materials, structures, or devices, and conditions that are fire hazards or that are hazardous to life, property or public welfare in any occupancy, structures or premises. The IFC requirements address the design, construction, installation, testing, and maintenance or removal of fire protection systems, including automatic sprinkler systems and fire alarm systems. Conditions that can affect the safety of fire fighters and emergency responders during emergency operations are also regulated by the IFC (Figure 1-9).
[Ref. 101.2]

FIGURE 1-9 Part of the IFC scope is the safety of first responders operating at the scene of an emergency.

The IFC, like other International Codes, is arranged and organized to follow sequential steps that generally occur during a plan review or inspection. The IFC is divided into seven different parts as shown in Table 1-1

TABLE 1-1 Major parts of the *International Fire Code*

Part	Title	Chapters
I	Administrative	1-2
II	General Safety Provisions	3-4
III	Building and Equipment Design Features	5-12
IV	Special Occupancies and Operations	20-39
V	Hazardous Materials	50-67
VI	Referenced Standards	80
VII	Appendices	Appendices A—N

There are several chapter numbers that are reserved for future use. These chapters allow for inclusion of new operations and processes as specific regulations are developed. When regulation of a new process is added to the code, it can be inserted into the appropriate "part" without renumbering all of the subsequent chapters. For example, in the 2015 IFC, Part IV included Chapters 20 through 37. Two new chapters have been added to the code and now Part IV includes Chapters 20 through 39; leaving Chapters 40 through 49 reserved for future use. This is intended to assist the code user by knowing that Chapter 57 regulates flammable and combustible liquids in this and future editions.

Several chapters in the IFC and IBC are coordinated to facilitate their use together. The IFC requirements for fire-resistance-rated construction, interior finish, fire protection systems, means of egress and fire safety during construction are directly correlated to the requirements in the IBC. Table 1-2 compares these chapters in the IFC and IBC.

TABLE 1-2 Correlated chapters in the International Fire and Building Codes

IFC Chapter	IBC Chapter
7 – Fire and Smoke Protection Features	7 – Fire and Smoke Protection Features
8 – Interior Finish, Decorative Materials and Furnishings	8 – Interior Finishes
9 – Fire Protection and Life Safety Systems	9 – Fire Protection and Life Safety Systems
10 – Means of Egress	10 – Means of Egress
33 – Fire Safety During Construction and Demolition	33 – Safeguards During Construction

Applicability of the IFC

Given its broad scope, the requirements in the IFC are applicable to both new construction and existing facilities. The IFC has design and construction requirements that apply to new buildings and processes, or when a change of use or change of occupancy occurs. Its requirements are intended to be used in conjunction with the requirements of the IBC. The IFC establishes requirements for the operations, processes and systems located indoors or outdoors, and maintenance of the building and equipment. [Ref. 101.2]

During an inspection, it is not uncommon to find a condition that is a violation of the IFC. If the violation has to do with the administration, operation or maintenance of the building, Section 102.2 requires that the violation be corrected using the requirements in effect at the time the violation was observed. In other words, the correction must comply with the code requirements applicable at the time the building is found in noncompliance with administrative, operational or maintenance provisions of the IFC. As an example, a fire inspector performing an inspection of a restaurant and nightclub finds that portable outdoor gas-fired heating appliances are being used inside an enclosed, canopy-covered area on a balcony (Figure 1-10). The jurisdiction has adopted the 2018 IFC. In this case, the inspector would cite the owner for violating Section 603.4.2.1.1, because it is violation of this section to use this type of heater inside of a canopy. [Ref. 102.2]

FIGURE 1-10 Two portable outdoor gas-fired heaters used inside a building are in violation of the operational requirements in the 2018 IFC, and the fire inspector would cite this installation as a violation of the requirements in IFC Section 603.4.2.1.1.

There is a different philosophy for application of the construction and design requirements in the IFC. While the operational, administrative and maintenance requirements are applicable to both new and existing buildings and facilities, the construction and design requirements are applicable to the following:

1. Structures, facilities and conditions that arise after the code is adopted.
2. Existing structures, facilities and conditions not legally in existence at the time the code is adopted.
3. Existing structures, facilities and conditions when required by the provisions in IFC Chapter 11.
4. Existing structures, facilities and conditions which, in the opinion of the fire code official, constitute a distinct hazard to life or property. [Ref. 102.1]

It must be understood that construction requirements referred to in Section 102.1 are not limited to building construction. In fact, most cases in the IFC are applicable to processes or equipment rather than structures or buildings. A trucking facility that has been in operation for several years decides it is time to provide on-site fueling. Without obtaining a permit, they install an above-ground tank and have a dispensing operation. This new tank and dispenser is considered construction and must comply with the current code when it is discovered.

Section 102.1 also applies to changes in "conditions." For example, a bottling company has been in operation since 1990, and the process was permitted and reviewed and complied with the code when it was installed. The owner decides to change the product they are bottling without any revision to the process line. This change in the process is considered a change in the "facility" or "conditions." The change in product would be treated as a new process and must meet the current code requirements

The requirements of the IFC are applicable for any conditions that arise after the date the code is adopted and for any new facilities or structures that are constructed. Item 1 of Section 102.1 states that if a developer wishes to construct a new apartment community and the jurisdiction has adopted the 2018 International Codes, then the site and building would be required to comply with the 2018 IFC.

It is not common, but it does happen that a process is changed or added to a facility without the benefit of an IBC building permit or IFC construction permit. When this situation is noticed by the inspector and there are violations found with the installation, this is considered an existing condition that is not legally in existence. It is considered "not legally in existence" because it did not comply with the code at the time of construction or installation. In other words, the installation never did comply, and therefore it is not legal. When this situation is identified, Item 2 of Section 102.1 states that since the work never complied with the code, it must now comply with the current code. Even if the process, equipment or construction has been in existence for a dozen years, it must comply with the current code when it is found.

Item 4 of Section 102.1 grants the fire code official the authority to retroactively enforce any provision in the 2018 IFC when, in his or her opinion, a situation presents a "distinct hazard to life or property." Declaring a building, process or use a "distinct hazard" requires a great deal of consideration, because once such a declaration has been made, the requirements must be equally enforced on all other similar buildings, processes or uses.

Such a declaration commonly results in legal challenges by the plaintiffs, due to the costs associated with the retroactive construction and renovating systems or features to comply with the IFC. In those instances where a fire code official considers a building, process or use a distinct hazard, such a decision should be discussed with the jurisdiction's attorney before making the declaration. In most cases, the retroactive application of a particular code requirement should consider the time required to obtain construction permits and to perform inspections, because immediate compliance may be difficult, if not impossible.

Retroactive construction requirements in the IFC

Item 3 of Section 102.1 refers to retroactive construction requirements for existing buildings. Typically, when a building is constructed in accordance with the code at the time of construction and the newer code changes, the building does not need to be modified to meet the newer code requirements. This situation is often referred to as a "legal nonconforming" or "existing nonconforming" building. In other words, it was legally constructed since it met the code at the time of construction, but it does not conform to the requirements in the current code. However, it has been found that many of the conditions that the codes previously allowed have since been revised because of serious consequences in fire conditions. Chapter 11 of the IFC stipulates construction requirements that are applicable to all existing buildings and facilities. These requirements specify the minimum fire safety requirements for existing buildings. Essentially, the ICC voting membership has made the determination that the items identified in Chapter 11 represent a "distinct hazard" to life safety, and these hazards must be mitigated.

For example, the code previously allowed open stairways in multistory buildings. Fire loss history shows that even when a fire occurred on the first floor, heated smoke traveled up the open stairway and fatalities occurred on floors above. For this reason, the code now requires stairways connecting more than two stories to be enclosed by fire-resistance-rated construction. Chapter 11 contains specific construction requirements that apply to existing buildings with identified distinct hazards. One of those requirements is to protect open stairways connecting more than two floors. However, Chapter 11 does not require compliance with the current code. For most construction features identified in Chapter 11, the owner is required to provide mitigation. In the example of the open stairway, the owner has the option to enclose the stairway or provide an automatic sprinkler system.

In occupancies that represent a high life-safety risk, such as an elementary school where the children must be under supervision [educational (Group E)], or a hospital where patients may not be capable of effecting self-rescue [institutional (Group I-2)], the IFC requires the retroactive installation of a fire alarm and fire detection systems to ensure that an unwanted fire is detected early in its growth. Other retroactive requirements address means of egress components, address numbers or letters on buildings, interior finish or the installation of automatic sprinkler systems in Group I-2 occupancies. The requirements in Chapter 11 are less restrictive than new construction requirements, but they are designed to mitigate specific life hazards in existing buildings.

> **You Should Know**
>
> Construction requirements in the IFC apply to new buildings and operations. Chapter 11 contains specific retroactive construction requirements for existing buildings with identified hazards.

Code Essentials

Change of occupancy is defined as

A change in the use of a building or a portion of a building that results in any of the following:
1. A change of occupancy classification.
2. A change from one group to another group within an occupancy classification.
3. Any change in use within a group for which there is a change in the application of the requirements of this code. •

Change of use or occupancy

The building official will issue a certificate of occupancy when a building is approved for occupancy. Every building constructed or renovated under the IBC receives a certificate of occupancy (Figure 1-11). In addition to the building's address, its building permit number and the adopted edition of the IBC used as the basis for its construction, the certificate of occupancy documents will include the following:

- The use and occupancy of the building,
- The building's construction type,
- The design occupant load,
- If an automatic sprinkler system is provided, whether the system is required, and
- Any special conditions or stipulations issued by the building official.

Fire code officials commonly use the certificate of occupancy during building inspections to verify the building's occupancy or use has not changed. If a building changes occupancy, for example, if a grocery store [mercantile (Group M occupancy)] is converted to a church with an occupant load of 550 [assembly (Group A-3 occupancy)] without review and approval of the jurisdiction, such an act can result in the code official issuing a stop use order. This occupancy change could require upgrading the building construction to increase its fire resistance, providing additional means of egress components, installing automatic sprinkler and fire alarm systems or upgrading the building's plumbing system. If the building is not designed for a

FIGURE 1-11 A Certificate of Occupancy is issued by the building official.

particular occupancy classification, it can potentially place the occupants in danger.

There are occasions where a change in use occurs, but the occupancy classification remains the same. A new Certificate of Occupancy is required, and the change is considered a change of occupancy if the change results in new or additional code requirements. Consider a Group B office building with multiple tenant spaces. A new tenant wants to establish an ambulatory care facility in this existing building. While the ambulatory care facility is classified as Group B, there are several specific code requirements applicable only to that use. When this tenant moves in, a building permit is required resulting from this change of use. When a change of use or change of occupancy occurs, the new occupancy is treated as new construction and must comply with the current code (Figure 1-12).

FIGURE 1-12 New building construction or renovation must comply with the requirements of the IFC and IBC.

The IFC generally prohibits a change of occupancy or use unless the change is done in conformance to the requirements of the IBC. The IFC allows changes in the use of a building provided it does not change the overall use or character. When approved by the fire code official, a building's use can change without conforming to all the requirements of the IBC, provided the new or proposed use is a less hazardous fire risk or life risk. For example, consider a Group S-1 storage occupancy storing rolled carpet. Rolled carpet has a very high heat release rate and requires a specialized automatic sprinkler system design. If the rolled carpet were removed and the warehouse were used for storing pallet loads of bottled water, the building could be reclassified as a Group S-2 occupancy, because the stored commodity has a much lower heat release rate. The fire code official could allow this use without requiring a change of occupancy since it is less hazardous than the use the building is designed to accommodate. [Ref. 102.3]

Historic buildings

In 1966, the U.S. Department of the Interior was assigned the responsibility of ensuring historic buildings were preserved under the National Historic Preservation Act. The legislation required each state to establish a historical building preservation office. As a result of this act, many communities also enacted their own local historic building preservation laws.

Historic buildings generally must be maintained in their original condition. Historic buildings may lack fire safety features normally required for new buildings having the same occupancy classification (Figure 1-13). These buildings also may not comply with means of egress requirements because they were constructed prior to the

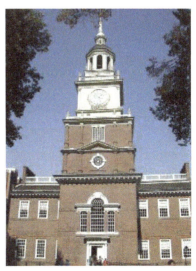

FIGURE 1-13 Historic buildings must comply with a fire protection plan approved by the fire code official.

development of fire and life safety design regulations in model codes and standards.

Unless the building is a distinct hazard, the IFC requires that historic structures be provided with fire protection and life safety features based on an approved fire protection plan. The criteria for developing a fire protection plan is contained in NFPA 914, *Code for Fire Protection of Historic Structures*. In some cases, the fire protection plan may need to be prepared as a performance-based design. In these instances, the design should be prepared based on the requirements in ICC *Performance Code® for Buildings and Facilities*. [Ref. 102.6, 1103.1.1]

Referenced codes and standards

The design, construction, testing and maintenance of a variety of systems or components are required by the IFC to comply with various technical standards. The IFC adopts over 200 different standards by reference in Chapter 80. Standards are formal documents that establish consistent and uniform technical or engineering criteria, methods and practices. When designing an automatic sprinkler system for the protection of a building, the IFC requires that it be designed in accordance with one of the three NFPA standards that govern the design of these systems. Essentially, the IFC will indicate where a fire protection system or safety component is required, and the referenced standards will provide the criteria for the design and installation of that system or component. The IFC also requires the evaluation of certain materials or components to be performed using standard test methods, which are definitive procedures for evaluating a product or component. The classification of a liquid as being either flammable or combustible is required by the IFC to be tested in conformance to one of four ASTM International standard tests to measure its boiling point and closed cup flash point temperatures. Compliance with the adopted technical standards or test methods is a requirement of complying with the IFC (Figure 1-14). [Ref. 102.7]

There are occasions where only a portion of a standard is referenced. In these situations, only those referenced provisions are applicable. For example, IFC Chapter 57, "Flammable and Combustible Liquids" contains over 40 references to specific requirements in NFPA 30, *Flammable and Combustible Liquids Code*, but never references the entire document. The IFC only intends to include those specific sections, and intentionally skips those portions that

FIGURE 1-14 The design of these petroleum storage tanks was required to comply with one of the atmospheric storage tank standards developed by the American Petroleum Institute.

either are already covered in the IFC or would create a conflict with the IFC. [Ref. 102.7.2]

There are also occasions where the IFC provisions are in conflict with a referenced standard. Most times, this conflict is intentional. The IFC specifically states that the code requirements supersede the requirement in the standard. Essentially, one can consider an IFC requirement to be an amendment to the standard. Many times, this different text in the IFC is more restrictive, but it could also be less restrictive. For example, NFPA 13R, Standard for the Installation of Sprinkler Systems in Low-Rise Residential Occupancies, allows for the elimination of fire sprinklers over balconies in multistory, multifamily dwellings. IFC Section 903.3.1.2.1 specifically requires fire sprinklers over balconies and decks in Type V construction. This requirement in the IFC is in conflict with NFPA 13R, but it must be complied with since the IFC requirement supersedes the standard requirement. [Ref. 102.7.1]

CHAPTER 2
Legal Aspects, Permits and Inspections

Fire codes intend to protect the health and safety of the public and emergency responders by establishing minimum requirements for the prevention of fires and explosions. To be effective, a fire code and its referenced standards must be adopted by a government jurisdiction and enforced by qualified officials appointed by the governing authority. Chapter 2 explains the process of adopting, amending and administering the *International Fire Code*.

CODE ADOPTION

The IFC and the other International Codes are commonly referred to as model codes. Model codes are nationally recognized regulations that address the design, construction, operation and maintenance of buildings, uses and hazards that are maintained and updated through an open and accessible code development process. The code development process relies on the participation of interested parties and individuals with experience in the areas of design, construction, manufacturing, logistics and code administration. The model codes are updated on 3-year cycles to recognize new and developing materials, technologies and methods of construction. Changes to the model codes are commonly in response to natural or technological disasters or significant events that result in human injury, death or the destruction of property. For additional information on the ICC Code Development Cycle, see Chapter 1.

Adoption of the IFC

A governmental jurisdiction must legally adopt the IFC for it to have the power of law. This is normally accomplished by an adopting ordinance that references the title and edition of the IFC (Figure 2-1). Generally, the legislation also includes the purpose, scope and its effective date. As part of the adopting legislation, the jurisdiction also provides information for insertion into code text including the name of the jurisdiction, limits for fines or imprisonment as a result of an individual or company violating the fire code, adopted appendices, permit fees and establishes land use limits for the above-ground storage of certain hazardous materials. [Ref. 101.1, 101.2.1, 110.4, 112.4]

Amending the IFC

Adoption of a model fire code offers the jurisdiction consistency and uniformity in performing plan reviews and inspections, as well as correlation among the adopted construction codes across different jurisdictional boundaries. Such uniformity benefits design professionals, builders and the regulated industries. While states, counties, local governments and similar jurisdictions do not have the resources to develop and maintain their own comprehensive fire protection and prevention codes, some jurisdictions have the ability to modify the model fire code through amendments placed in the adopting ordinance. Excessive local amendments to adopted model codes are contrary to the goals of consistency and can offset the advantages and legal defensibility of nationally recognized standards. Some states prohibit amending any construction codes, and the local jurisdiction must enforce what the state legislative or authorized administrative body adopts.

> **ORDINANCE NO. 2018-009**
>
> **ADOPTION OF THE 2018 INTERNATIONAL FIRE CODE; ESTABLISHING PERMITS FEES, FINES; DECLARATION OF FIRE CODE VIOLATIONS AS A CRIMINAL VIOLATION OF THE CITY CODE; ESTABLISHING LAND USE BOUNDARIES FOR CERTAIN FLAMMABLE HAZARDOUS MATERIALS**
>
> An ordinance of the City adopting the 2018 edition of the International Fire Code, regulating and governing the safeguarding of life and property from fire and explosion hazards arising from the storage, handling and use of hazardous substances, materials and devices, and from conditions hazardous to life or property in the occupancy of buildings and premises in the City; providing for the issuance of permits and collection of fees therefore; repealing Ordinance No. 2015-006 of the City and all other ordinances and parts of ordinances in conflict therewith.
>
> The City Council does ordain as follows:
>
> **Section 1.** That Chapter 17 of the Ordinance Code of the City is hereby repealed, and a new Chapter 17 is hereby added to read as follows:
>
> **Section 2.** That a certain document, three (3) copies of which are on file in the office of the City Clerk of the City, being marked and designated as the International Fire Code, 2018 edition, including Appendix Chapters B, C, D, J and N as published by the International Code Council, as modified and amended by this ordinance is hereby adopted as the Fire Code of the City, regulating and governing the safeguarding of life and property from fire and explosion hazards arising from the storage, handling and use of hazardous substances, materials and devices, and from conditions hazardous to life or property in the occupancy of buildings and premises as herein provided; providing for the issuance of permits and collection of fees therefore; and each and all of the regulations, provisions, penalties, conditions and terms of said Fire Code on file in the office of the City with the additions, insertions, deletions and modifications in this ordinance, are hereby referred to as the City Fire Code.
>
> **Section 3.** That the following sections of the City Fire Code are hereby revised:
> Section 101.1 is amended to read: Insert: [NAME OF JURISDICTION]
> Section 110.4 is amended to read: Insert: [OFFENSE, DOLLAR AMOUNT, NUMBER OF DAYS]
> Section 112.4 is amended to read: Insert: [DOLLAR AMOUNT IN TWO LOCATIONS]
> Section 1103.5.3 is amended to read: Insert: [DATE BY WHICH SPRINKLER SYSTEM MUST BE INSTALLED]
>
> **Section 4.** That the geographic limits referred to in certain sections of the City Fire Code are hereby established as follows:
> Section 5704.2.9.6.1: geographic limits in which the storage of Class I and Class II liquids in above-ground tanks outside of buildings is prohibited: [JURISDICTION TO SPECIFY]
> Section 5706.2.4.4: geographic limits in which the storage of Class I and Class II liquids in above-ground tanks is prohibited: [JURISDICTION TO SPECIFY]
> Section 5806.2: geographic limits in which the storage of flammable cryogenic fluids in stationary containers is prohibited: [JURISDICTION TO SPECIFY]
> Section 6104.2: geographic limits in which the storage of liquefied petroleum gas is restricted for the protection of heavily populated or congested areas: [JURISDICTION TO SPECIFY]
>
> **Section 5.** That if any section, subsection, sentence, clause or phrase of this legislation is, for any reason, held to be unconstitutional, such decision shall not affect the validity of the remaining portions of this ordinance. The City hereby declares that it would have passed this law, and each section, subsection, clause or phrase thereof, irrespective of the fact that any one or more sections, subsections, sentences, clauses and phrases be declared unconstitutional.
>
> **Section 6.** That nothing in this legislation or in the Fire Code hereby adopted shall be construed to affect any suit or proceeding impending in any court, or any rights acquired, or liability incurred, or any cause or causes of action acquired or existing, under any act or ordinance hereby repealed as cited in Section 1 of this ordinance; nor shall any just or legal right or remedy of any character be lost, impaired or affected by this legislation.
>
> **Section 7.** That the City Clerk of the City is hereby ordered and directed to cause this legislation to be published. [An additional provision may be required to direct the number of times the legislation is to be published and to specify that it is to be in a newspaper in general circulation. Posting may also be required.]
>
> **Section 8.** That this law and the rules, regulations, provisions, requirements, orders and matters established and adopted hereby shall take effect and be in full force and effect [TIME PERIOD] from and after the date of its final passage and adoption.

FIGURE 2-1 Sample ordinance adopting the IFC

While the model codes anticipate location and climate differences, local amendments to the code are generally influenced by unique characteristics and conditions such as engine or ladder company response times or staffing, the available water supply, land use limitations or community topography and geography. There also may be considerations of community politics, customs or traditions, regional practices and other local or state laws. Lengthy amendments to the IFC are not uncommon. Amendments must be legally instituted through the adopting ordinance and laws of the jurisdiction. Otherwise, actions to enforce requirements not contained in the IFC or to waive requirements in the model code will not be in accordance with the law.

Appendices

The appendices are developed in the same manner as the body of the model IFC. However, appendices commonly are judged as being outside the scope and purpose of the model code. Appendices offer supplemental information, alternative methods or recommended procedures. The information may serve as a guideline or an example of a recommended practice. In instances where an appendix serves as a source of information, such as Appendix G which offers equivalent weight and volume values for cryogenic fluids, the jurisdiction is not required to adopt the code text, but can use the technical content for plan reviews and inspections. If a jurisdiction wishes to adopt and enforce an appendix chapter, for example Appendix B on Fire-flow Requirements for Buildings, then the appendix must be specifically identified in the adopting ordinance (see Section 2 in Figure 2-1). Appendix chapters or portions of the appendix that gain acceptance over time are sometimes moved into the main body of the model code. [Ref. 101.2.1]

> **You Should Know**
>
> The appendices are not an enforceable part of the IFC unless they are specifically listed in the adopting regulation.

Local and state laws

The requirements in the IFC are not meant to nullify any local, state or federal law, and in some instances, those provisions will supersede the requirements in the fire code. For example, Exception 4 of Section 5001.1 exempts off-site transportation of hazardous materials from the IFC when the transportation is in accordance with the U.S. Department of Transportation (DOT) requirements. The Hazardous Materials Transportation Act pre-empts the IFC because DOT has authority over the packaging and transportation of hazardous materials, regardless of the transportation mode (Figure 2-2). If a fire code official attempts to regulate the transportation of hazardous materials, DOT can and has obtained a federal injunction prohibiting the jurisdiction from enforcing IFC requirements.

FIGURE 2-2 Off-site transportation of hazardous materials is regulated by the U.S. DOT, and its requirements supersede those in the IFC.

Careful consideration should be given to the correlation of IFC with other codes and laws of the jurisdiction during the adoption process. Zoning ordinances may have more restrictive land use limitations than those set forth in the IFC. Other local laws that may impact the IFC include those regulating public streets, traffic calming, storm water management, fire hydrant locations and spacing or fire pro-

FIGURE 2-3 Sample engineer's seal

> **You Should Know**
>
> "The fire code official is hereby authorized to enforce the provisions of this code and shall have the authority to render interpretations of this code, and to adopt policies, procedures, rules and regulations in order to clarify the application of its provisions."
> International Fire Code Section 104.1

> **Code Essentials**
>
> **Fire code official duties**
> - Enforce the IFC
> - Review construction documents and permit applications
> - Issue permits, notices and orders
> - Conduct inspections
> - Maintain records
> - Investigate the cause and origin of unwanted fires
> - Control the scene of emergencies
>
> **Fire code official authority**
> - Make interpretations
> - Adopt policies, procedures and regulations
> - Approve alternative methds, materials and modifications
> - Investigate fires, explosions or other hazardous conditions

tection water supplies from water wells. State laws may supersede local codes related to certain hazardous materials used as fuel gases, underground storage tanks and licensing of contractors who design and install fire protection systems. State law also dictates when a registered design professional, including a registered professional engineer or licensed architect, is required to seal design drawings and specifications, and state law sets the licensing requirements for these design professionals (Figure 2-3). [Ref. 102.11]

AUTHORITY

The IFC establishes a Department of Fire Prevention, commonly referred to as a Fire Prevention Bureau or Office of the Fire Marshal, and designates the individual in charge of the implementation, administration and enforcement of the code as the fire code official. The appointing authority of the jurisdiction appoints the fire code official. The fire code official is authorized to designate individuals as deputy fire code officials who may perform plan reviews and inspections as well as other technical and administrative staff. The position of fire code official demands skills, knowledge and abilities not only to fulfill the duties but also to maintain and elevate the credibility of the Department of Fire Prevention and the fire department in the eyes of the public. [Ref. 103.1, 103.2, 103.3]

Authority and duties of the fire code official

The IFC charges the fire code official with enforcing the provisions of the fire code and assigns broad authority and discretion to do so. With discretion comes the responsibility to make decisions in keeping with the intent of the IFC. Conversely, the fire code official has no authority to require more than the code stipulates. The IFC authorizes the fire code official to develop policies, procedures and regulations to clarify the application of the provisions. These policies, procedures and regulations must be within the spirit and intent of the code, and often they are used to explain how the code official intends to enforce the code (Figure 2-4). [Ref. 104.1]

To effectively perform the prescribed duties, the fire code official must have an understanding of the legal aspects of code enforcement. While given broad authority for enforcement, including but not limited to issuing a stop use order or disconnection of building utilities, the fire code official must also recognize the rights of due process afforded to the public. Equally important to the Department of Fire Prevention in securing safe buildings, facilities and uses for the community and its emergency responders is to build the public trust through communication, respect and fairness so that the organization is viewed as a resource rather than an adversary. [Ref. 110.4.1, 112.1, 113.1]

Standard No. 503.6-1		
Kern County Fire Department Fire Prevention 	**Electronically Controlled Access Gates**	*Fire Marshal* Revised: January 1, 2016 Date: January 1, 2008

This Standard is promulgated in accordance with Section 104.1 of the Kern County Fire Code and is an official interpretation of Section 503.6 of the Kern County Fire Code.

BACKGROUND

Firefighting access must be provided to facilities and structures. Business and home owners often desire to install security gates across access roads, fire lanes, driveways, and entrances to communities. The consideration needs to be made for the ability to provide security in addition to the need to provide access in an emergency.

Section 503.6 of the Kern County Fire Code reads as follows:

> **503.6 Security gates.** The installation of security gates across a fire apparatus access road shall be approved by the fire code official. Where security gates are installed, they shall have an approved means of emergency operation. The security gates and the emergency operation shall be maintained operational at all times. Electric gate operators, where provided, shall be listed in accordance with UL 325. Gates intended for automatic operation shall be designed, constructed and installed to comply with the requirements of ASTM F 2200.

REQUIREMENTS

This Standard provides methods for emergency services agencies to obtain access when the access road is blocked by a security gate. The following items must be complied with:

1. A permit is required to install an electronically controlled access gate which obstructs a fire department access road.
2. Plans and specifications for access gates shall be submitted to the Fire Prevention Office for review and approval prior to construction.
3. Emergency access through the security gate will be provided by either of the following methods:
   a. Electronically controlled gates shall be provided with a 3M Opticom receiver system.
   b. Electronically controlled gates shall be provided with security guard, or staffed, on a 24-hour basis.
   c. A Knox Key Operated Dual Switch (KS3503) shall be installed at each security gate.
      i. The Knox Key Switch must have a switch designated for FIRE and POLICE.
      ii. The Knox Key Switch shall allow for emergency override of any electrical devices.
      iii. The Knox Key Switch shall have a "normal" and an "emergency" position. The Key Switch shall be installed so that in the "emergency" position all gates open and remain open for emergency access until the switch is returned to the "normal" position.
      iv. The Knox Key Switch shall be located to be easily accessible and visible.
      v. The location of Knox Key Switches are subject to approval of the Fire Department.
4. Security gates shall be capable of being manually operated or shall be provided with battery back-up in the event of power failure which shall cause the gate to remain in the open position until power is restored.
5. Security gates shall be maintained operational at all times. When electronically controlled gates are out of service, they shall be secured in the open position until repairs are complete. Repairs shall be in accordance with original specifications.
6. Security gates shall have a minimum clear opening at least as wide as the required access road width.
7. An operational test shall be requested by the installer and witnessed by the fire department prior to placing the system into operation to establish that the final installation complies with the requirements, the specified design, and is functioning properly.

FIGURE 2-4 Fire department interpretation, or policy, on electronically controlled access gates gates *(Courtesy of Kern County [CA] Fire Department)*

Technical assistance

In certain instances, a permit applicant will submit design drawings and specifications to the Department of Fire Prevention that involve a building or system that is complicated or technically challenging. This is not uncommon when dealing with buildings or processes that store, handle or use hazardous materials, the design of automatic sprinkler and standpipe systems for high-rise buildings or specialized automatic sprinkler systems protecting high-piled combustible storage. To ensure the commissioned design complies with the requirements of the IFC and its adopted standards, the fire code allows the fire code official to obtain technical assistance (Figure 2-5). The fire code official is authorized to use a third party to review the design drawings and specifications to verify that the design complies with the IFC. The cost of the review and any reports is the responsibility of the permit applicant. The individual performing the review and preparing the report and opinion must be approved by the fire code official. A fire protection engineer or chemical engineer may be requested by the jurisdiction to perform the review and to provide technical guidance to the fire code official, but a registered design professional is not stipulated by the code. It is critical that the technical assistance is provided by someone who is qualified in the subject matter to be addressed in the report. The qualified person will provide a report to the fire code official. Review and acceptance of the report, and the responsibility for the final approval of the permit rests with the fire code official. [Ref. 104.7.2]

FIGURE 2-5 The fire code official can utilize the technical assistance provision to ensure that the design and construction of difficult or complicated industrial processes such as this chemical blending operation meet the fire code requirements.

Alternative materials and methods

The IFC is specific in its intention not to exclude the use of any material or method of construction not specifically prescribed by the code, subject to the approval of the fire code official. Given the increasing pace at which technology advances, new and innovative materials and construction methods are being constantly introduced into the market. The fire code official has an obligation, as instructed by the code, to approve such alternatives when it is demonstrated by the design professional or permit applicant that the proposed material or construction method offers equivalent quality, strength, effectiveness, fire resistance, durability and safety when compared to the IFC requirements. In many instances, the alternative material or method will exceed the requirements of the fire code and provide a greater level of safety. When evaluating an alternative material or method, it is important to ensure that the proposed method or material meets the intent of the applicable IFC provisions.

One method to assist code officials reviewing alternative materials and methods is the International Code Council's Evaluation Service (ICC-ES) reports (Figure 2-6). ICC-ES reports are a resource that is available to code officials to verify that the performance of a system, construction method or component equals the code requirements. In the absence of ICC-ES reports or sufficient data or docu-

ICC-ES Evaluation Report

ESR-5000

Issued January 2018
This report is subject to renewal January 2019.

www.icc-es.org | (800) 423-6587 | (562) 699-0543 | *A Subsidiary of the International Code Council®*

DIVISION: 07 00 00—THERMAL AND MOISTURE PROTECTION
Section: 07 30 05—Roofing Felt and Underlayment

REPORT HOLDER:

ACME UNDERLAYMENTS UNLIMITED
52380 FLOWER STREET
CHICO, MONTANA 43820
(808) 664-1512
www.underlaymentunlimited.com

EVALUATION SUBJECT:

UU 100 UNDERLAYMENT FOR ASPHALT SHINGLE ROOF COVERINGS IN SEVERE CLIMATE AREAS

1.0 EVALUATION SCOPE

Compliance with the following codes:

- 2018 and 2015 *International Building Code®* (IBC)
- 2018 and 2015 *International Residential Code®* (IRC)

Property evaluated:

Ice barrier

2.0 USES

UU 100 is a self-adhering, rubberized asphalt membrane, complying with ASTM D1970, that is used over plywood substrates as ice barriers as specified in Chapter 15 of the IBC and Chapter 9 of the IRC.

3.0 DESCRIPTION

The UU 100 has a granule surfacing. The membrane has a silicone-treated release paper on the back that is removed prior to attachment to plywood sheathing. The membrane is a minimum of 0.040 inch (1.02 mm) thick and is supplied in rolls 36 inches (914 mm) wide and 66.7 feet (20.3 m) long.

4.0 INSTALLATION

Installation of the UU 100 membrane must comply with this report and the manufacturer's published installation instructions. The manufacturer's published installation instructions and this report must be available at the jobsite at all times during the installation.

Prior to application of the membrane, the deck surface must be free of frost, dust and dirt, loose nails and other protrusions. Damaged sheathing must be replaced. Installation is limited to plywood substrates. The membrane is designed for applications when the ambient air temperature is above 40°F (4.4°C).

Vertical ends and horizontal edges must be overlapped a minimum, respectively, of 6 inches (152 mm) and 3½ inches (89 mm). Horizontal edge overlaps must run with the flow of water in a shingling effect. A minimum of two layers of the membrane must be applied, starting at the lower edge (eave) of the roof, and extend a minimum of 24 inches (610 mm) inside the exterior wall line of the building. Final coverage width must comply with the code.

Installation of the roof covering can proceed immediately following application of the membrane. The membrane must be covered by an approved roof covering as soon as possible. For reroofing applications, the same procedures apply after removal of the old roof covering and roofing felts to expose the plywood roof deck.

5.0 CONDITIONS OF USE

The UU 100 membrane described in this report comply with, or are suitable alternatives to what is specified in, those codes listed in Section 1.0 of this report, subject to the following conditions:

5.1 Installation must comply with this report and the manufacturer's published installation instructions. In the event of a conflict between the manufacturer's published installation instructions and this report, this report governs.

5.2 Installation is limited to use on plywood substrates on structures located in areas where non-classified roof coverings are permitted.

5.3 Installation is limited to use with roof coverings that are mechanically fastened through the underlayment to the sheathing or rafters.

5.4 Installation is limited to roofs with ventilated attic spaces, in accordance with the requirements of the applicable code.

6.0 EVIDENCE SUBMITTED

6.1 Data in accordance with the ICC-ES Acceptance Criteria for Self-adhered Roof Underlayment for Use as an Ice Barrier in Severe Climate Areas (AC48), dated February 2012 (editorially revised December 2015).

6.2 Reports of testing in accordance with ASTM D1970.

7.0 IDENTIFICATION

The membrane is identified by labels on the rolls or packaging, displaying the Acme Underlayments Unlimited's name and address, the product name, and the evaluation report number ESR-5000.

FIGURE 2-6 Sample ICC Evaluation Services Report

mentation, the IFC authorizes the fire code official to require a test of the system, component or construction method to verify it meets the requirements of the IFC and its adopted standards. Testing must be performed by an approved agency and the test method requires the fire code official's approval. In instances where the fire code official does not feel qualified to make a determination as to which test method(s) to use or to review the findings of a test report, the code official can choose to request technical assistance. [Ref. 104.9.2]

Authority at fires and other emergencies

The scope of the IFC includes conditions affecting the safety of fire fighters and emergency responders during emergency operations. The IFC grants the fire code official, the fire chief or the incident commander at the scene of a fire or other emergency to control and direct the incident scene for the protection of life and property or take any other actions that are necessary in the reasonable performance of duty (Figure 2-7). This can include disconnecting building utilities or limiting or prohibiting the movement of people and vehicles that are not authorized at the incident scene. These requirements allow those in charge to deploy fire-fighting resources based on the size and magnitude of the incident. Consider a large commercial building fire. It may be necessary to lay water supply lines from hydrants located one to two blocks from the fire scene. The IFC requirements authorize fire department officials with the ability to block roads and streets and to limit the areas in which the public or media have access. [Ref. 104.11]

FIGURE 2-7 The IFC grants the officer in charge of controlling a structure fire complete control of the scene, including limiting access to vehicles and the public.

A fire or medical emergency can emotionally impact family members and friends. In some cases, people can become extremely agitated, especially if the emergency concerns a close family member, friend or family pet. People can lose focus that the emergency responders are trained professionals and that the methods for controlling fires or patient treatment may not be understood by those observing the emergency.

To ensure the care of patients is not compromised and to protect the safety of emergency responders treating the individual or managing the emergency, the IFC allows the incident commander, fire chief or fire code official to barricade the scene. The IFC also grants emergency personnel the authority to have individuals detained who may be obstructing operations if they disobey a lawful command from fire or police department personnel (Figure 2-8). [Ref. 104.11.1, 104.11.2]

FIGURE 2-8 The fire chief, fire code official or officer in charge of the incident has the authority given in the IFC to secure the emergency scene and order the removal of individuals who obstruct operations.

PERMITS

A permit is a written document issued by a code official that legally authorizes an individual or business to conduct certain businesses, services or construction in accordance with the requirements of the jurisdiction's adopted codes. The IFC requires the fire code official to issue permits to perform certain hazardous operations or activities and for the construction or alteration of fire protection systems and processes storing and handling hazardous materials or highly flammable materials. Under specified circumstances, the IFC exempts certain hazardous operations and activities and allows limited repair to fire protection systems; however, any work that is exempt from a permit still must comply with the applicable IFC requirements. [Ref. 105.1.1, 105.3]

If the construction, alteration or operation of the system or process is not performed and maintained in conformance to the IFC requirements in effect when the permit was issued, the fire code official can revoke the permit. Permit revocation essentially is a stop use order and the activities must cease until violations are corrected and compliance is demonstrated to the fire code official. [Ref. 105.3.6, 105.3.8]

Operational and construction permits

The IFC authorizes the fire code official to issue permits for certain hazardous operations, which are termed "operational permits," and permits for the construction or alteration of fire protection systems and equipment or systems designed for the storage and use of highly flammable materials or hazardous materials, identified as "construction permits." [Ref. 105.1.2]

The 2018 IFC requires an operational permit for 50 hazardous processes or activities that are regulated by the code. Operational permits can be issued for a prescribed time with an expiration date. The date could be based on the length of operation, such as a temporary amusement building, or based on an annual or biannual renewal. Operational permits may also be issued without an expiration date. During routine inspections, the permitted operation must remain in compliance with the code or the permit is revoked. The jurisdiction will need to establish a policy or procedure for the time duration permitted for operational permits. **[Ref. 105.3.1, 105.5, 105.6]**

The IFC requires 25 different construction permits. In addition to the construction or alteration of fire protection systems, including automatic sprinkler systems, standpipes and private fire protection fire hydrants and water distribution piping, construction permits are required for temporary membrane structures and tents. Permits are also required for systems storing and using compressed and liquefied compressed gases, cryogenic fluids, flammable and combustible liquids and other hazardous materials. Construction permits usually expire when the work has been completed and approved by the code official. **[Ref. 105.7]**

It is very common for a particular address or regulated use to require multiple operational permits. For example, a factory manufacturing welded bicycle frames may need operational permits for the welding operation (hot work permit), the spray painting of the completed frames (spraying or dipping permit) and the storage of the bicycle in cardboard boxes stored on steel racks awaiting shipment (high-piled combustible storage permit). When multiple permits are required for one facility, the code allows the permits to be consolidated into a single permit that covers all of the permitted operations. **[Ref. 105.1.3]**

There are many times when a permit is issued for construction, and when the construction is completed another permit is issued for the continued operation (Figure 2-9). For example, a typical motor vehicle fuel-dispensing station will store, handle and dispense flammable and combustible liquids. The liquids will be stored in above-ground or underground storage tanks that will deliver these liquids, via a pump and piping network, to motor vehicle fuel dispensers. Company personnel and the general public may have access to the dispensers. In such a case, a construction permit is required for the installation of the tanks, piping, pumps and dispensers, and an operational permit is required for the storage, handling, use and dispensing of flammable and combustible liquids. A construction permit is not required for the maintenance of the installed system.

FIGURE 2-9 This above-ground fuel dispensing operation requires IFC construction and operational permits issued and approved by the fire code official.

Construction documents

Drawings and specifications must accompany any construction permit application and be of sufficient detail and clarity to verify compliance with the code. In some cases, the code will require a site plan showing all new and existing structures with distances to buildings on the same property, building openings and property lines. The extent of construction documents varies with the complexity and scope of the project.

Construction documents are defined by the IFC as "the written, graphic and pictorial documents prepared or assembled for describing the design, location and physical characteristics of the elements of the project necessary for obtaining a permit." Construction documents must comply with the preparation and submittal requirements in Section 105.4. When required by the jurisdiction or state laws, the construction documents must be prepared by a registered design professional, such as a licensed architect or engineer. The fire code official can waive this requirement when it is demonstrated that the nature of the work does not require the services of a registered design professional. [Ref. 202, 105.4, 105.4.1]

Fire protection system drawings and supporting calculations must be prepared in accordance with the applicable NFPA standards referenced in Chapter 9. These plans must also comply with any specific requirements set forth in Chapter 9, such as fire protection of wooden decks and balconies in wood frame residential (Group R) buildings or the use of quick-response sprinklers or residential sprinklers in certain institutional occupancies. The fire code places the responsibility for the preparation of the construction documents on the permit applicant and requires each permit application to be complete. [Ref. 105.4.3]

During large or extended construction projects, the registered design professional may want to use a phased design approach in which construction documents are submitted based on various project milestones. In many buildings, the use of these *design-build* construction methods has proven to provide major cost savings because the building is designed as each of the major elements is being constructed. Phased approval of construction documents is allowed by the IFC; however, the designer and the contractor are required to assume any risks associated with improper or incomplete construction methods or installations, and as the project concludes, the fire code official may withhold any approvals until all of the code requirements are satisfied. [Ref. 105.4.4.1]

Permit application

Before a permit can be issued, the owner or an authorized agent must apply in such form and detail as required by the fire code official. When required by the IFC or the fire code official, construction documents also must accompany the permit application. Because the degree and level of information can vary between operational and construction permits, the jurisdiction should establish a clear procedure and policy for the minimum information required for each permit that is issued by the Department of Fire Prevention (Figure 2-10).

Application for Permit
Request for Service

Kern County Fire Department

Office of the Fire Marshal – Fire Prevention
2820 M St. ~ Bakersfield, CA 93301
Telephone (661) 391-3310 ~ Fax (661) 636-0466 or (661) 636-0467
Website: kerncountyfire.org Email: fireprevention@kerncountyfire.org

Date: ____________________ *To be completed by permit Applicant (PLEASE TYPE)*

Project Name: ____________________

Project Street Address: ____________________

City: ____________________

☐ **Fire Alarm** ___ # of Devices
☐ **LPG Tank(s)** ___ # of tanks
☐ **Special Effects**
☐ **Hood Suppression** Fire Protection Systems
☐ **Spray Booth** spraying & dipping
☐ **Plan Review**
☐ **Automatic Sprinkler** Remodel ☐ or New ☐
☐ **Tents/Canopies** complete Tent Handout
☐ **Knox Box**
☐ **Flammable Liquids** _____ # of tanks
☐ **Explosives** Storage or Use
☐ **Other** ____________
☐ **Flow Test**
Name of Water District:
Contact # for Water District:

Description of project: ____________________

If additional space is needed please attach to this Application

Will any hazardous materials be stored: ☐ Yes ☐ No

If yes, include a list of containers and quantities: ____________________

Customer/Applicant Information

Applicant Name: ____________________ Phone #: ____________

Contractor/Company: ____________________ Phone #: ____________

Contractor Mailing Address: ____________________

City: ____________________ State: ________ Zip: ____________

Contractor License #: ____________ Class: ________ Fax #: ____________

Contact Person: ____________________ Cell #: ____________

Email Address: ____________________

Permit applications may be submitted in person or mailed to:
Kern County Fire Department
2820 M St.
Bakersfield, CA 93301
Attn: Fire Prevention--Permits

Visa and MasterCard Accepted
Please include a check made payable to: **Kern County Fire Department**

Permit Number:____________________

____
Signature

KCFD 200 (0717) ***Also submit site plans, construction plans and specifications***

FIGURE 2-10 Plan review application for a construction permit *(Courtesy of Kern County [CA] Fire Department)*

The fire code official is required by the code to examine each permit application to determine if the proposed activity or construction complies with the applicable code requirements. If the submitted information or construction documents do not comply with the requirements in the fire code or its referenced standards, the fire code official can reject the submittal and must provide a written reason to the applicant for its denial. [Ref. 105.2.1, 105.2.4, 105.4.1.1]

In many cases, a plans examiner may find that the application and its supporting documentation are acceptable; however, certain portions or features of the design may need to be modified to meet the intent or letter of the fire code. In such cases, the fire code official can issue an approval of the permit but stipulate certain conditions or requirements that must be satisfied. The code official cannot waive or set aside any other provisions of the IFC, the jurisdiction's adopted construction codes or other laws. [Ref. 105.3]

The fire code official is authorized to perform an inspection of any building, process or system before an operational permit is issued. This inspection can be used to establish any operational constraints or limits as well as to determine if any other operational permits are required (Figures 2-11 and 2-12). [Ref. 105.2.2]

FIGURE 2-11 The fire code official can inspect the facility before an operational permit is issued.

| City Fire Department
Fire Prevention Division | Construction Permit |

| PERMIT NUMBER: 2018-SPR-156 | PERMIT DATE: June 15, 2018 |
| DESCRIPTION: Installation of sprinkler system – New Const. | EXPIRATION DATE: June 15, 2019 |

| ADDRESS: 2613 Moffitt Way | BUILDING/SUITE: Bldg. C |

| SCOPE OF WORK: Wet-pipe sprinkler system – High-piled combustible storage |

CONDITIONS: All City Regulations and the 2018 International Fire Code shall apply.

Systems are subject to field inspection – to schedule an inspection, call (661) 555-3473.

This permit shall expire twelve (12) months from the date of issue.

This permit is not transferrable.

SCOPE: Installation of a wet-pipe automatic sprinkler system using existing risers for the protection of high-piled combustible storage. Stored product is classified as a Class IV commodity – Group B plastic. Storage is double row racks with a maximum height of 17 feet. Design areas 1, 2, 3, and 5 protected using Central Model K17-231 (K=16.8) pendent sprinklers. Discharge density of 0.44 GPM/sq. ft. Design area of 3,000 sq. ft. Design basis is FM Global Data Sheet 8-9. Design area 4 is the flammable liquid pump room and is classified as Extra Hazard Group 2.

See the plan review requirement for special inspection requirements for pipe hangers and supports for the 8" feed mains.

DESIGN DEMANDS:
Design Area 1 Demand (BOR): 1528 GPM @ 33.3 PSI (Sprinkler K=16.8)
Design Area 2 Demand (BOR): 1528 GPM @ 42.6 PSI (Sprinkler K=16.8)
Design Area 3 Demand (BOR): 1672 GPM @ 49.1 PSI (Sprinkler K=16.8)
Design Area 4 Demand (BOR): 618 GPM @ 49.7 PSI (Sprinkler K=8.0)
Design Area 5 Demand (BOR): 1476 GPM @ 32.1 PSI (Sprinkler K=16.8)

| COMPANY NAME: KHS Fire Sprinklers |

| COMPANY ADDRESS: 11223 Main Street, City, CA 99999 |

| RESPONSIBLE PERSON: William Purveyor, NICET D-08911 |

Approved plans shall be available and permits shall be posted on-site at all times during which work authorized hereby is in progress. Progress is defined as the time from which site work begins until the time of final Fire Department approval. Inspections shall not be conducted if approved plans and permits are not on-site. Violation of this requirement is subject to a Stop Work order and civil/criminal sanctions as allowed by the code.

FIGURE 2-12 A fire code construction permit

Fees

Jurisdictions generally establish a fee schedule at a level to offset the costs of providing services to the public, including administration, plan reviews and inspections. Many jurisdictions choose not to charge fees for routine inspections performed by fire companies; however, if a fire company inspection finds a permit that should be referred to the Department of Fire Prevention, then fees for these services can be assessed. Permit fees are commonly charged for the renewal of operational permits. An inspection should be performed prior to the renewal and fees could cover administration and inspection. Permit fees could also be based on the total value of the construction project or on inspection history for various occupancy classes or equipment. Whatever the method selected, the fire code official should develop an equitable and consistent procedure for assessing and refunding fees related to the permit process. The IFC does not specify that fees are required or the appropriate dollar amount for the fees; however, if the jurisdiction requires permit fees, the code states that the fee must be paid before a permit is issued. [Ref. 106]

INSPECTIONS

Inspections are an important part of confirming and verifying fire code compliance. For new construction, additions or tenant improvements, inspections are performed to confirm that fire protection systems are installed in accordance with the approved design presented in the construction documents and in conformance to the referenced standards. In many jurisdictions, a fire inspection is a requirement for licensing of day care and health care occupancies. Inspections to verify compliance with the fire code are typically required by building code officials before a certificate of occupancy is issued.

In the case of the IFC, an inspection is required before an operational permit can be issued. Additionally, construction permits require inspections throughout the construction or installation process. Equipment, processes and facilities requiring a construction under the IFC cannot be utilized or occupied until the necessary inspections have been made and the code requirements have been met. [Ref. 105.3.3, 107.2]

Right of entry

IFC Section 104.3 establishes provisions that authorize the fire department staff to conduct inspections. Because fire inspections can be required for any property, vehicle or vessel, permission to perform the inspection must first be obtained from the property owner, tenant or an individual authorized to allow entry onto the property. A property owner can refuse an inspector's entry into a building or onto a site, as this right is protected by the Fourth Amendment to the U.S. Constitution:

> The right of the people to be secure in their persons, houses, papers and effects against unreasonable searches and seizures, shall not be violated, and no Warrants shall be issued, but upon probable cause supported by Oath or affirmation and particularly describing the place to be searched and the persons or things being seized.

In 1967, the U.S. Supreme Court issued a ruling involving a fire inspection in the case of *See v. The City of Seattle* (387 U.S. 541, 87 S.Ct. 1737, 18 L.Ed.2d 943). The appellant (Norman See) sought reversal of his conviction for refusing to allow a representative of the City of Seattle Fire Department to enter and inspect his locked commercial warehouse without a warrant and without probable cause to believe that a violation of any municipal ordinance existed. The inspection was conducted as part of a routine, periodic city-wide canvass to obtain compliance with Seattle's Fire Code. After he refused the inspector access, Mr. See was arrested and charged with violating the fire code. He was convicted and given a suspended fine of $100, despite his claim that the authority to inspect provision of the fire code, if interpreted to authorize this warrantless inspection of his warehouse, would violate his rights under the Fourth and Fourteenth Amendments. The Supreme Court agreed and issued its ruling in Mr. See's favor, which has since governed how code officials and inspectors must initiate inspections.

To satisfy the requirements of the Fourth Amendment and the rulings of the Supreme Court Justices, fire code officials must respect the requirements of the law and perform certain steps before initiating an inspection on or into a protected area. A protected area is not visible from any public view or areas where privacy is expected. In almost every case where the requirements of the IFC are followed, entry must be granted by the individual responsible for the building or premises so the inspection can be performed (Figure 2-13):

FIGURE 2-13 Before performing an inspection, the fire code official must obtain consent from the owner or a responsible party.

1. Identify yourself, the basis and reason for inspection. The jurisdiction should issue an official method of identifying code officials. **[Ref. 104.4]**
2. Obtain permission and consent from a responsible individual with the business, building or site. Note that the courts have ruled against third-party consent to conduct an inspection when it was granted by the child of an owner or an employee. Therefore, it is important to obtain either oral or written consent from an individual who is responsible for the site or building.
3. If the inspection is for the purpose of verifying compliance with a construction or operational permit required by the IFC, inform the individual that this is the basis for inspection.
4. The inspection shall occur during reasonable times. The court has ruled that "reasonable times" means during normal working hours. If the building is vacant, the fire code official will need to locate the owner to schedule an inspection.

5. In certain cases, a business will request a copy of the legal basis for an inspection. If asked, either cite or provide a copy of IFC Sections 107 and 104.3.
6. Once consent is granted, the inspection can proceed. [Ref. 104.3]

If access is denied, the inspector should document the circumstances, time and date of the incident, and contact the jurisdiction's legal counsel for advice on how to proceed. The code allows the inspector to obtain a warrant, commonly referred to as an inspection warrant or administrative warrant. This type of warrant allows for the inspection to occur based on the normal inspection process of the jurisdiction. Contrary to a search warrant, probable cause or evidence of a crime is not needed to obtain an inspection warrant. [Ref. 104.3.1]

Liability

Fire code officials conducting code enforcement inspections often are anxious about their potential liability exposure if they make a mistake or overlook a potential problem. News reports describing multimillion dollar settlements, the fire fighters' unfamiliarity with the code enforcement process and an inherent suspicion of the legal profession combine to raise the apprehension level. "Can I get sued?" "Will I be held liable?" "Who will protect me?" are common questions for the inexperienced code official as well as fire suppression personnel who may be assigned to code enforcement.

Any person can sue another for some alleged error or insult. A code official can be held liable or found negligent only if it is established after all the facts have been heard at a trial and a jury renders a verdict. Fortunately, there are many protections built into the American jurisprudence system that make the likelihood of being found liable a remote possibility as long as the fire code official or fire inspector is acting within the scope of their authority and training, referred to as a "standard of care" or "due care." There are protections in the IFC itself that limit a code official's or fire inspector's liability. The fear of being sued should never interfere with a jurisdiction's commitment to perform code enforcement for the safety of fire fighters and the public.

The laws regarding liability are incredibly complex and differ in every state; a detailed discussion is beyond the scope of this book. In some states, government officials are protected from lawsuits arising from any work they perform in their governmental roles (they are said to possess "governmental immunity"), but this principle has been eroded in many states in recent years. If a code official or inspector is concerned about liability exposure while performing inspections or code enforcement, the individual should discuss those issues with legal counsel. It might be a city or county attorney, or a professional who is on retainer to provide legal advice and consultation.

> **You Should Know**
>
> To be found liable of negligence, the following four criteria must be proven:
> - There was a "duty"
> - There was a "breach of duty"
> - There was a "proximate cause"
> - There were damages as a result.

In very general terms, to prove someone is liable of negligence, the person bringing the charge (the "complainant" or "plaintiff") must prove the defendant is negligent of four things: duty, breach of duty, cause and damages.[1]

"Duty" describes conditions where the defendant had a legal obligation to perform a specific task, and often that specific task has to be provided for a certain group or "class" of persons. For example, in many states, fire fighters and emergency medical services personnel have a legal obligation (a "duty") to report child or elder abuse to law enforcement. These same first responders may not have a duty to report abuse to someone who does not meet the legal definition of a "child" or "elder" citizen. The duty or obligation to perform building and fire code enforcement varies by jurisdiction.

"Breach of duty" means the defendant failed to fulfill the obligation. If an inspector has a legal obligation to report electrical hazards to the local electrical inspector and fails to do so, he may be accused of breaching that duty.

"Cause" means that whatever the alleged code violation or oversight might have been, it has to have led to the problem for which the defendant is accused. Additionally, the cause must be "proximate," in other words, it must be the direct cause of the damages. Imagine that a fire fighter has a duty to inspect sprinkler fire department connections and fails to check one on a commercial inspection. Later, it is determined that the connection had been badly damaged by a delivery truck and the fire department could not use it during a fire. The complainant might argue successfully that had the fire department been able to use the fire department connection, the fire's outcome might have been different. There is a link (a "cause") between the failure to inspect the connection and the inability to use it.

Finally, there must be "damages," or harm. Without damage, there is no liability. In the previous example, if the fire was controlled by a single sprinkler and extinguished by hand lines without spreading beyond the area of origin, the plaintiff would be hard pressed to claim there was any damage caused by the broken fire department connection because it did not play a part in the incident.

If the plaintiff cannot prove all four elements (duty, breach of duty, cause and damages) the defendant likely will not be found negligent or liable.

The IFC provides another level of protection against liability. The IFC states that "the fire code official or any subordinate shall not be liable for costs in an action, suit or proceeding that is instituted in pursuance of the provisions of this code; and any officer of the department of fire prevention, acting in good faith and without malice, shall be free from liability for acts performed under any of its provisions or by reason of any act or omission in the performance of official duties in connection therewith." This protection extends to

---

1. Owen, David G. *Hofstra Law Review*, "The Five Elements of Negligence," Volume 35, No. 4, Summer 2007, p. 1672.

the fire code official, fire inspector, fire fighter or company officer conducting inspections within the scope of his or her work. The IFC also requires the jurisdiction employing the fire code official to provide legal defense and to cover the costs of any settlement that might result from a trial. [Ref. 103.4.1]

The IFC is also clear that any approval resulting from an inspection does not allow a violation of the code. This means that even though an inspection is completed and no items were noted in violation, but the inspector missed an item which is in violation of the code, the item in violation does not suddenly become approved. The fact that the inspection was approved is not an approval of the violation. The item is still in violation and the owner is responsible for correction. At the next inspection, the item in violation may be found and must be corrected. [Ref. 107.4]

Testing and operation

The fire codes place the responsibilities for the proper installation of any required fire protection, life safety or hazardous materials storage and use systems in the hands of the registered design professional or the installing contractor. Once these systems are approved by the fire code official, it is the owner's or tenant's responsibility to ensure that they are maintained in accordance with the fire code requirements, including its referenced standards. [Ref. 108.1]

Many of the systems specified by the IBC and IFC are required to be inspected or tested annually, or on a specified schedule. These inspections are the responsibility of the owner or tenant, and most businesses will retain a contractor to perform required inspections and any maintenance. Many of the inspections can be accomplished by the owner or the owner's representative; however, repairs and maintenance must be accomplished by either a contractor or qualified individual. It is the responsibility of the owner or tenant to maintain these records so they can be reviewed by the fire code official. In some jurisdictions, inspection reports for testing of fire protection equipment and systems are submitted to the fire code official for review. [Ref. 108.3] During these system inspections and tests, the contractor or fire code official may identify fire code violations that require correction (Figure 2-14). The IFC requires that the necessary corrections be performed and that the fire code official reinspect or witness retesting of the corrected area of concern to verify compliance. [Ref. 108.2.1]

FIGURE 2-14 The fire code official is authorized to perform as many inspections as necessary to verify that systems are properly installed and maintained.

Unsafe buildings

If, as a result of a natural or technological emergency, or as a result of the lack of maintenance or changing a building's use, a fire code official finds a building or

FIGURE 2-15 This underground gasoline storage tank was not adequately secured and floated above ground after extended rains saturated the soil. Because of the threat of fire and unauthorized release of hazardous materials, the fire code official issued orders for the removal of the stored flammable liquid and the damaged storage tank.

equipment that represents a serious fire or life safety threat, the IFC grants broad authority to require the necessary corrections to bring it into code compliance (Figure 2-15). The scope of this authority extends to building systems or any system regulated by the fire code. Any actions necessary to satisfy the requirements of the IFC that involve repair or upgrades to building structural components or building demolition require the approval of the building code official. When beginning the process of correcting unsafe buildings or conditions, a written notice to the owner, tenant or responsible party is required. **[Ref. 111.1]**

If the fire code official finds that the violation constitutes a hazardous condition that presents an imminent danger to the building occupants, the code official is authorized to require the partial or complete evacuation of a building and can prohibit re-entry into the building. **[Ref. 111.2]**

Stop work order

In some instances, individuals or businesses will perform work that is regulated by the IFC without obtaining the required operational or construction permits. Under certain circumstances, the continued work or operation of equipment represents dangerous or unsafe conditions to employees or personnel who may have access to the site or building. In these cases, the fire code authorizes the fire code official to issue a stop work order (Figure 2-16). A stop work order can also be issued when an inspector arrives to conduct an inspection of the underground piping for the automatic sprinkler system and finds that the piping is already backfilled and covered. The stop work order will not allow further work to continue until the work already completed has been uncovered and properly inspected. **[Ref. 112.1]**

When a fire code official issues a stop work order, it must be in writing, must explain the basis for stopping the work, and must state the conditions under which the cited work is

---

DEPARTMENT OF FIRE PREVENTION

STOP WORK ORDER

_____123 Main Street, Big City, CA_____
Site Address

The address listed above is in violation of the City Fire Code, 2018 IFC, Section 105.7 as amended and adopted by reference in Section 17.32.091 of the Municipal Ordinance Code, **installing an aboveground fuel tank with a Permit.**

ALL WORK SHALL BE STOPPED and shall not continue until approval is granted by the Fire Code Official and all necessary permits are obtained.

For information contact the Department of Fire Prevention at (888) 555-1234.

WARNING

This notice is to be removed only under the authority of the Fire Code Official.

Date Posted: _____   By: _____
                                      Fire Code Official

PENALTY IF REMOVED WITHOUT AUTHORIZATION—Signs, tags or seals posted or affixed by the Fire Code Official shall not be mutilated, destroyed or tampered with or removed without authorization from the fire code official (IFC 110.3.4) and are subject to a $250.00 fine.

To file an appeal to this order, send notice in writing to Big City Board of Appeals, 121 Downtown Avenue, Big City, CA.

FIGURE 2-16 Stop work order

allowed to resume. Unless otherwise stipulated by the fire code official, all stop work orders are an immediate compliance order—when issued, the activity affected by the legal order must stop immediately. Failing to comply with the stop work order is generally treated by most jurisdictions as a criminal violation and can be subject to fines. [Ref. 112.2, 112.4]

BOARD OF APPEALS

The IFC administrative provisions create authority and duties for the fire code official but intend that actions in enforcing the code be reasonable while protecting due process rights for the regulated community. Under the provisions of the IFC, a property or business owner has the right to legally challenge the code interpretations of a fire code official or inspector. Any person or organization that has a material interest in the fire code official's decision may apply for a hearing or review to the Board of Appeals (Figure 2-17). The jurisdiction's governing body, such as its board of directors or city council, appoints the board members, who are qualified by experience and training to hear and rule on interpretations issued by the fire code official. [Ref. 109.1, Appendix A]

The IFC limits the basis for appeals to matters pertaining to code requirements. The appellant must claim the fire code official has erred in interpreting the code or has wrongly applied a code section. The IFC does not allow the filing of an appeal to seek a variance or waiver, nor does it grant the Board of Appeals the authority to waive code requirements. [Ref. 109.2]

Courts have ruled that permit and inspection fee requirements cannot be waived because they are costs associated with an inspection program. However, a permit applicant still has the right to appeal.

FIGURE 2-17 Application to the Board of Appeals *(Courtesy of Phoenix [AZ] Fire Department)*

PART II

General Safety Requirements

Chapter 3: General Precautions against Fire

Chapter 4: Emergency Planning and Preparedness

CHAPTER 3

General Precautions against Fire

The general safety requirements in the IFC were developed to control a wide variety of fire safety concerns that may not need additional clarification or the level of detail that might be found in other chapters.

Chapter 3 covers combustible waste materials (such as wood, paper and plastics) and sources of ignition. Ignition sources include mechanical, chemical, electrical or optical energy. The chapter also addresses topics such as fire safety issues related to vacant premises, fueled equipment and mobile food preparation vehicles. Vacant premises can be a major fire hazard to communities because if they are not adequately secured they may be used for criminal activity or as illegal and substandard housing. "Fueled equipment" includes motorcycles, mopeds, lawn-care equipment and portable cooking equipment. Fueled equipment can be found in a variety of buildings and work sites, and represents another fire hazard because of the fuels used and the common indoor use of the equipment. While not a structure or building, mobile food preparation vehicles contain inherent fire and life safety hazards, such as cooking and storage of compressed or liquefied gases. Chapter 3 also contains requirements addressing hazards to fire fighters and fire-fighting operations.

COMBUSTIBLE MATERIALS

Combustible materials are natural or synthetic materials that can be ignited and support combustion. Combustible materials in the context of this chapter and IFC Chapter 3 are not combustible metals or flammable solids—these are hazardous materials that are regulated by other provisions in the fire code. Materials regulated by IFC Chapter 3 generally are organic materials such as sawn wood, dimensional lumber, waste paper or cardboard and baled cotton or paper, or synthetic materials such as plastics, fabrics or composite materials. Combustible materials are always solids and will have varying sizes and densities. The smaller the surface area of a combustible material and the lighter its density, the more easily it is ignited. The orientation of the combustible material, the strength of the ignition source and other variables can influence the ignition of combustible materials.

FIGURE 3-1 Ignition of combustible materials beneath the egress stairs can eliminate the stairs as an escape path, and therefore storage in that location is a violation of the IFC.

The fire code recognizes that combustible materials are an important part of businesses and industries. The combustible material requirements in IFC Chapter 3 address the orderly storage of these materials, locating the materials away from ignition sources and, if the storage is indoors, separating the combustible materials from means of egress components and concealed spaces where they could accelerate the rate at which an unwanted fire grows and spreads. Orderly storage can slow the rate of fire spread, which benefits fire fighters in the event that the materials are ignited (Figure 3-1). **[Ref. 315]**

While it is not within the scope of this chapter, fire code officials should understand that storage of many combustible materials over 12 feet in height inside of buildings introduces the potential for a fire that will exhibit a much faster growth rate when compared to the same materials stored at or near the floor level. Such storage can be found in many warehouses and mercantile occupancies and is required to comply with the requirements in IFC Chapter 32. Chapter 14 in this book introduces the reader to the hazards of high-piled combustible storage.

Combustible waste

When combustible materials become "waste," the IFC takes a more aggressive approach: the materials must be removed and disposed of in a controlled manner. For most combustible wastes, the IFC requires that they be placed in noncombustible waste containers or plastic containers formulated from chemicals that reduce the amount of heat it releases if ignited. When materials are placed in bulk trash receptacles (dumpsters), the fire code requires the dumpsters be located at least 5 feet from combustible construction, wall openings and combustible roof eaves (Figure 3-2). Because of land use limitations, it is very common to place dumpsters inside

Code Essentials

The IFC requirements for combustible materials depend on whether the material is used as a material or goods or if it is waste material.

Combustible material must be stored in an orderly manner, away from ignition sources and in locations that do not disrupt the means of egress.

Combustible waste must be located in approved waste receptacles. The IFC has specific requirements for dumpsters located indoors and outdoors.

FIGURE 3-2 The IFC requires separation of outdoor dumpsters from buildings to limit the likelihood of the dumpster igniting an exposure building.

of buildings. In such instances, the room housing the dumpster is required to be protected by an automatic sprinkler system. Sprinkler protection is not required when the dumpster is located in a building constructed of noncombustible, fire-resistive materials and used exclusively for dumpster or trash container storage. [Ref. 304.3.4]

Outdoor pallet storage

Transportation of products on pallets is a common occurrence in many different businesses. Whether the product is shipped off-site, or simply relocated within the facility, pallets are a useful tool providing quick access and mobility. A pallet adds a moderate fire load to the product it is carrying. When a pallet is not being used, it is considered an idle pallet. When idle pallets are stacked, the fire load and accompanying hazard increase dramatically. This occurs because each piece of wood in a wooden pallet is able to burn on all sides, and the typical construction of a pallet allows adequate air to reach all surfaces of the wood. So rather than burning from the outside in, like a stack of 2 × 4's each piece of wood can be burning at the same time (Figure 3-3). [Ref. 315.7.5]

FIGURE 3-3 Outdoor storage of idle pallets can pose a significant fire exposure.

The outdoor storage of pallets is specifically regulated because a fire in the pallet storage area creates an enormous amount of heat and can easily impact any exposures. The separation of pallet piles from buildings, property lines and other pallet piles is contingent on the type of pallet and the construction of the building (Figure 3-4). Pallets that are of wood and plastic pallets that are labeled in accordance with UL 2335, Fire Tests of Storage Pallets, or FM 4996, Approval Standard for Classification of Pallets and Other Material Handling Products as Equivalent to Wood Pallets, are all treated as a group with regard to spacing and pile size requirements. Plastic pallets not in compliance with either standard are treated as a separate group since the heat release rate is considerably higher than wood pallets. [Ref. 315.7]

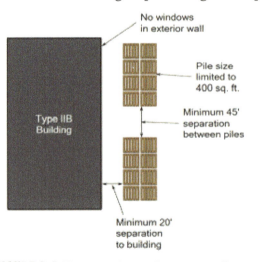

FIGURE 3-4 These outdoor pallet storage piles are separated from the building based on piles of more than 200 pallets. Note: if there were any windows in the exterior building wall, the separation distance would change from 20 feet to 90 feet.

IGNITION SOURCES

Controls for ignition sources are dictated in several chapters of the IFC, including specific requirements for electrical equipment and hot work involving brazing, oxygen-acetylene cutting and welding. IFC Chapter 3 contains general requirements to address separating uses and activities involving potential sources of open flames from combustible materials. The provisions require adequate separation between open flames and combustible materials, open-flame warning devices such as road flares and negligent burning of combustible vegetation and materials. Cooking, decoration, theatrical or construction activities are regulated elsewhere in Chapter 3. [Ref. 305]

An open-ended requirement for the control of ignition sources and unwanted fires is included in IFC Section 305.5. This section states that when situations, uses or processes have repeatedly caused fires, the cause of the fire must be mitigated. This section can be used to address systems or situations that are not specifically regulated in the code, but are creating an unsafe situation resulting in repeat fires. There are no specific requirements to address the hazard other than the requirement to modify the situation to prevent further fires. [Ref. 305.5]

OPEN FLAMES

The IFC allows the use of open flames for theatrical performances, food preparation, religious ceremonies, decoration and paint removal. Open flames are prohibited in sleeping units of Group R-2 dormitories and for cooking on combustible balconies of Group R-1 and R-2 occupancies unless the balconies are protected by an automatic sprinkler system. Under very limited conditions, open flames are permitted in assembly (Group A) occupancies. The IFC requires an operational permit for using open flames in assembly areas, dining areas of restaurants and drinking establishments. [Ref. 105.6.34, 308.1.4, 308.4.1]

When open flames are used for decorations, the fuel source cannot be liquefied petroleum gas or a liquid with a flash point temperature less than 140°F (Figure 3-5). If the device contains more than 8 ounces of fuel, it must be designed to be self-extinguishing and have a limited rate of fuel release if it is tipped over. The decorative flame source must be adequately secured and located so it is not an ignition source of interior finishes such as shades or curtains or other combustible materials. [Ref. 308.3.1]

FIGURE 3-5 An open-flame decorative device

FIGURE 3-6 Open flames used to prepare food and beverages are regulated by the IFC.

Open flames are commonly used in the table-side preparation of food and beverages (Figure 3-6). These activities commonly occur in assembly occupancies such as restaurants and nightclubs; therefore, the use of open flames in an occupancy with a large occupant load requires close supervision and detailed regulations. The IFC limits the volume of liquid that can be dispensed to 1 ounce or less per serving and limits the container volume to 1 quart. The activity must have a controlled flame height and is limited to the immediate area where the food is prepared for consumers. Flaming foods and beverages may not be carried through the restaurant or nightclub. The person who prepares the flaming food or beverage is required to have a wet cloth towel to extinguish the flame in the event of an emergency. [Ref. 308.1.8]

VACANT PREMISES

Vacant buildings can present a significant fire threat to a community. These buildings can be used by transients for housing or for illegal activities. The building itself can be made unsafe by the theft of plumbing and electrical components manufactured from copper, brass or other valuable materials. Thieves will open walls and shafts to remove these building materials, creating vertical paths for fire spread. To limit the risk of unwanted fires, the IFC has requirements for fire safety in vacant buildings.

Buildings that are vacated can be demolished by the jurisdiction (Figure 3-7). In many communities, the jurisdiction may place a lien on the property to recover the demolition costs. Demolition generally occurs when a building is continuously used for illegal activities, is structurally unsafe, presents a fire hazard or is a public nuisance. In other cases, the building may be secured and eventually reoccupied or even renovated. In such cases, the securing of the building or its renovation must comply with the IBC, the *International Property Maintenance Code* and the IFC (Figure 3-8). [Ref. 311.1.1]

> ## Code Essentials
>
> Vacant buildings must be safeguarded to limit the potential for vandalism or acts of arson. The IFC requires that the building's fire protection system be maintained and requires the removal of any combustible materials and hazardous materials. The fire code official is authorized to require placarding of a vacant building to identify fire-fighting hazards.

FIGURE 3-7 This vacant building needs significant repair or it will be demolished. *(Courtesy of New Orleans [LA] Fire Department)*

Safeguarding a building requires that openings into the structure, such as doors and windows, are protected from unauthorized entry (Figure 3-9). Whenever possible, fire protection systems should be maintained in service; however, this can be difficult especially in cold weather environments that can freeze water in wet-pipe sprinkler or standpipe systems or in hot, humid environments that can cause corrosion in electronic components installed in fire alarm control units and smoke detectors. In these cases, the fire code official can permit the system to be disabled, provided that combustible materials and hazardous materials are removed from the building and the building's location in relation to other exposure buildings does not represent a fire hazard. In all cases, any fire-resistance-rated construction and assemblies must be maintained in vacant buildings to limit the spread of fire. [Ref. 311.2]

FIGURE 3-8 A vacant building that is not properly safeguarded.

The IFC authorizes the fire code official to placard unsafe buildings to warn fire fighters of interior hazards (Figure 3-10). The placard is used to indicate if a structure is safe to enter during fire-fighting operations or to indicate certain structural and life safety hazards to fire fighters. For example, the placard could indicate that the roof is open, fire escapes are unsafe, holes exist in the flooring or stairways are unsafe. Placards are required on all sides of a building and at entry doorways. The IFC dictates the minimum size and symbols required on the placard. [Ref. 311.5]

FIGURE 3-9 A safeguarded vacant building

You Should Know

IFC Chapter 3 stipulates the minimum precautions against fire. The requirements address ignition sources, open burning and recreational fires, open flames, powered industrial trucks and equipment, smoking, vacant premises, indoor displays and miscellaneous combustible materials storage. ●

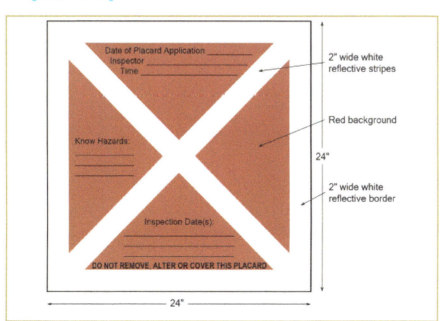

FIGURE 3-10 Fire-fighter building warning placard

FIGURE 3-11 A vehicle display inside a covered mall building

INDOOR DISPLAYS

Indoor displays of goods, vehicles or exhibitions must be located and arranged so they are not an obstruction to the means of egress. The IFC prohibits the indoor display of fireworks, flammable and combustible liquids, liquefied compressed flammable gases, oxidizers, agricultural goods and pyroxylin plastics in main exit access aisles, malls and corridors or within 5 feet of any means of egress opening if the fire code official believes a fire could prevent or otherwise obstruct egress. [Ref. 314.3]

Vehicle displays inside of buildings must be adequately safeguarded to limit the amount of fuel and ignition sources (Figure 3-11). The IFC requires that such displays limit the amount of fuel to 5 gallons or one-quarter of the tank volume, whichever is smaller, and that the fuel tank fill opening is sealed and the batteries are disconnected. [Ref. 314.4]

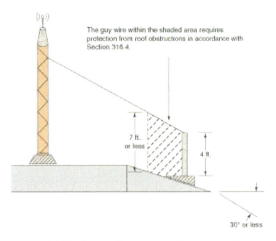

FIGURE 3-12 Obstructions on rooftops must be protected so as not to create a hazard to fire-fighting operations.

HAZARDS TO FIRE FIGHTERS

Each year, thousands of fire fighters are either hurt or fatally injured during fire-fighting operations. To help to limit these tragedies, the IFC addresses some potential hazards inside buildings and on the roofs of buildings. Doors providing access to shaftways must be identified where there is no landing on the opposite side of the door. Trapdoors to basements or scuttle covers need to be closed when not in use. Obstructions on rooftops that could "clothesline" a fire fighter during roof operations need to be protected (Figure 3-12). [Ref. 316]

FIGURE 3-13 This grass-covered roof provides a park-like area for recreation.

ROOFTOP GARDENS AND LANDSCAPED ROOFS

Planting vegetation on building rooftops is becoming a common occurrence. The vegetation provides several functions. It provides a pleasing area for recreational activities or meetings, as well as a significant level of thermal insulation for the building (Figure 3-13). The vegetation also creates several concerns for the inspector and fire fighter. Roof ventilation is no longer a viable solution when the roof is covered with several feet of soil, and dying or dead vegetation creates a fire hazard.

The IFC regulates rooftop gardens and landscaped roofs. The landscaped portion of the roof is limited in size to a maximum area of 15,625 square feet with a maximum dimension of 125 feet. The limitation on size provides a fire break and a location for roof ventilation. Additional landscaped areas can be provided as long as they are separated by a distance of not less than 6 feet. This separation must consist of a Class A-rated roof system (Figure 3-14). IFC Section 905.3.8 requires that when a standpipe is installed in the building, it must be usable and available for the landscaped roof. [Ref. 317.2, 317.3, 905.3.8]

FIGURE 3-14 Landscaping on the roof must be maintained in a safe condition and allow fire-fighting access.

A landscape maintenance plan can be required that would include trimming of trees and shrubs, watering schedule and a list of vegetation species. The plan should provide for the removal of dead and decaying material at least twice a year. [Ref. 317.4]

MOBILE FOOD PREPARATION VEHICLES

Mobile food preparation vehicles equipped with appliances that produce smoke or grease-laden vapors are regulated in IFC Section 319. These vehicles present the same hazards as commercial cooking operations in a restaurant. The difference is that these vehicles are mobile and customers normally are not inside the vehicle. The customers being outside provides them with an easier escape, but the occupants of the vehicle typically have a single exit door available at one end of the vehicle. There have been incidents within the past several years resulting in injuries and fatalities. In order to regulate these vehicles, an operational permit is required in IFC Section 105.6.30. [Ref. 319.1]

> **You Should Know**
>
> The new requirements for mobile food preparation vehicles only applies to those vehicles producing smoke or grease-laden vapors.

Mobile food preparation vehicles are required to be provided with an exhaust hood over the cooking appliances generating the smoke or grease-laden vapors. The exhaust hood must be equipped with a fire-extinguishing system and a portable fire extinguisher must be provided. [Ref. 319.3, 319.4]

The fuel gas system supplying the cooking appliances is regulated whether the fuel is LP-gas or CNG. With either gas, the on-board quantities are limited. Cooking oil is also limited and cooking oil storage tanks must be listed. [Ref. 319.7]

FIGURE 3-15 Mobile food-preparation vehicle

CHAPTER 4
Emergency Planning and Preparedness

Planning for an emergency is an important aspect in the overall fire and life safety of building occupants and includes prompt notification of emergency services, clearly documenting a building's or site's fire safety plans, training of employees in how to respond, assigning fire watch or crowd managers to assist in evacuation and providing the necessary information to first responders so they can safely and quickly mitigate the hazard. Buildings or facilities storing and handling hazardous materials must provide a mechanism so the hazards of these materials are understood by individuals who work with these chemicals. Chapter 4 presents the IFC provisions that address emergency planning and preparedness, including notification of emergency forces, public assemblies and events, fire safety plans, emergency evacuation drills and employee training and response to emergencies.

EMERGENCY FORCES NOTIFICATION

Prompt notification of emergency responders is essential to controlling an incident involving a fire or unauthorized discharge of hazardous materials (Figure 4-1). In the event of an emergency, the IFC requires immediate notification of the fire department and implementation of appropriate emergency plans and procedures by employees and occupants of the building. The fire code prohibits any delay in reporting a fire as well as reporting any false alarm. [Ref. 401.3]

FIGURE 4-1 Immediate notification of a fire or hazardous material release will result in a response by the fire department.

PUBLIC ASSEMBLIES AND EVENTS

Public assemblies and events can occur inside buildings or outdoors. Depending on the nature of the event and the number of persons estimated to be in attendance, the fire code official is granted authority to ensure that fire and life safety is maintained. One means of ensuring that safety is maintained is the use of fire watch personnel. Depending on the event, fire watch personnel can be any approved individual, or the jurisdiction may require the use of fire department personnel. The IFC authorizes the fire code official to require fire watch personnel based on the anticipated number of persons or the nature of the performance, exhibition or activity. Fire watch personnel are responsible to maintain watch for any unwanted fire and to extinguish any incipient fire, ensure means of egress paths and openings are maintained, and assist in evacuating people from the event. [Ref. 403.12.1]

Some events can have a major impact on a community, such as the arrival of dignitaries or an annual community festival. In other than assembly and educational occupancies, the IFC allows the fire code official to develop or prescribe a public safety plan for such large events to address issues such as fire apparatus access, emergency medical response, law enforcement, weather monitoring and similar issues (Figure 4-2). [Ref. 403.12.2]

> **Code Essentials**
>
> A public safety plan would provide
> 1. Emergency vehicle access
> 2. Fire protection
> 3. Emergency escape routes
> 4. Public assembly areas
> 5. Persons to direct attendees and vehicles
> 6. Vendors and concessions
> 7. The need for on-site law enforcement
> 8. The need for on-site fire services
> 9. The need for on-site emergency medical services
> 10. The need for a weather monitoring person

CROWD MANAGERS

Another method of providing for the safety of the occupants in assembly occupancies or outdoor events is through the use of crowd managers. The IFC requires that crowd managers are provided for events with a large

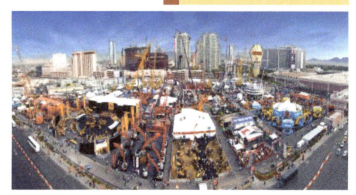

FIGURE 4-2 A large community event such as this trade-show event may require the preparation of a public safety plan to protect participants. *(Courtesy of Ed Kaminski, Clark County Building and Fire Prevention Services)*

> **Code Essentials**
>
> Crowd managers are required where the attendance is
> 1. More than 500 inside a building
> 2. More than 1,000 inside an assembly used exclusively for religious worship
> 3. More than 1,000 outdoors.

attendance. The number of crowd managers is based on the occupant load, with one crowd manager required for every 250 people. The difference between crowd managers and fire watch personnel is that the crowd managers normally function as facility employees and react as crowd managers when an incident happens, whereas the fire watch personnel are only providing fire watch duties. For example, at a large banquet hall at a hotel, the wait staff could also be trained crowd managers. During normal operations, the wait staff performs table prep and customer service. However, when an incident occurs, the wait staff's functions change to crowd managers, and they each have an assigned duty. In this example, the wait staff could be assigned to take the customers at their assigned tables to the closest exit and outside to a predesignated meeting area. The crowd manager function is designed to provide for the safety of the occupants by having the crowd managers take charge and lead the customers to safety. This could even be to a secondary exit when the closest exit is unusable. [Ref. 403.12.3]

The IFC requires that crowd managers are appropriately trained and that the training is approved. The term "approved" allows the fire code official full interpretation and full authority to determine what will or will not be approved. There are established training programs available for crowd manager training. One of these training programs is endorsed by the ICC and the National Association of State Fire Marshals. This training is provided through the Fire Marshal Support Services, LLC. Training is available via the internet and can be found through the ICC Preferred Provider Program, or at www.crowdmanager.com. This training program provides education on all aspects of crowd manager duties and tests the applicant's knowledge. If the applicant is successful, he or she will receive a certification card as crowd manager. [Ref. 403.12.3.2]

FIRE SAFETY AND EVACUATION PLANS

Certain buildings require fire safety and evacuation plans. The IFC requirements are based on the building's occupancy classification, the occupant load or a particular building use. The plan is maintained and updated by the building owner when staff assigned to perform fire safety functions changes, the building occupancy changes or a physical change to the interior architecture occurs. The plan must be available to all employees. Fire safety plans are not required for Group S and U occupancies. [Ref. 404.2]

A fire safety and evacuation plan is required for the following occupancies and buildings:
- Group A (Assembly) other than Group A occupancies used exclusively for purposes of religious worship that have an occupant load less than 2,000

- Group B (Business) buildings having an occupant load of 500 or more persons or more than 100 persons above or below the lowest level of exit discharge
- Group B ambulatory care facilities
- Group E (Educational)
- Group F (Factory/Industrial) buildings having an occupant load of 500 or more persons or more than 100 persons above or below the lowest level of exit discharge
- Group F pallet manufacturing and recycling facilities
- Group H (Hazardous)
- Group I (Institutional)
- Group M (Mercantile) buildings having an occupant load of 500 or more persons or more than 100 persons above or below the lowest level of exit discharge
- Group R-1 (Residential, transient)
- Group R-2 college and university buildings
- Group R-4 (Residential, custodial care)
- High-rise buildings
- Covered malls exceeding 50,000 square feet in aggregate floor area
- Open mall buildings exceeding 50,000 square feet in aggregate area within the perimeter line
- Underground buildings
- Buildings with an atrium and containing an occupancy of Group A, E or M
- Buildings utilizing occupant evacuation elevators
- Buildings containing high-piled combustible storage with more than 500,000 square feet of Class I–IV commodities or more than 300,000 square feet of high-hazard commodities. [Ref. 403]

The evacuation plan documents whether the building will be evacuated in its entirety or partially, such as in a high-rise building where selective evacuation may be appropriate, or a Group I-2 hospital where relocation may be appropriate. The plan identifies key personnel who need to remain to operate critical equipment (such as process control or emergency power systems), operations requiring fire watch or crowd managers, procedures for determining that the partial or complete evacuation has been completed and all persons are accounted for, the primary and alternate procedures for notifying the occupants of the evacuation and the method for notifying the fire department. If the building is equipped with an emergency voice/alarm communications system, the primary and any secondary messages that will be announced must be documented and approved. [Ref. 404.2.1]

EMERGENCY EVACUATION DRILLS

With the exception of Group H, M, S and U occupancies, all other IFC regulated occupancies require one or more annual emergency evacuation drills. An emergency evacuation drill is defined as "an exercise performed to train staff and occupants and to evaluate their efficiency and effectiveness in carrying out emergency evacuation procedures." An emergency evacuation drill is an exercise to familiarize employees or occupants with the building's fire safety and evacuation plan and is conducted to ensure that the plan is properly implemented (Figure 4-3). Additionally, a new design for evacuating high-rise buildings includes the option to install occupant evacuation elevators. Traditionally, people have learned to avoid elevators in times of emergency. However, new elevator designs and safety components have been combined to allow the use of elevators for the transport of people who are unable to exit via the stairways. Specific training is needed for the employees in these high-rise buildings to be cognizant of the elevator operations, where such elevators are provided. Deficiencies or limitations of the plan are commonly identified during these drills, and these findings should be used to improve the fire safety and evacuation plan. [Ref. 405.1]

FIGURE 4-3 An evacuation path of a multistory Group B office building

The frequency of emergency evacuation drills and the required participants are set forth in IFC Table 405.2 (Table 4-1). This table establishes the number of times within a year that emergency evacuation drills are required and specifies who is required to participate. For example, IFC Section 403.6 requires a fire safety and evacuation plan for Group F occupancies with an occupant load of 500 or more persons or more than 100 persons located above or below the lowest level of exit discharge. Under the provisions in Table 405.2, plant employees must participate in an emergency evacuation drill annually. [Ref. Table 405.2]

When an evacuation drill is planned, the event should be scheduled to occur at different times and under varying conditions to simulate unusual conditions that could occur. Records documenting particular aspects of the drill, such as the person who supervised the event, its date and time, the method of initiating the event, the number of persons evacuated and the time to complete the evacuation must be maintained for review by the fire code official. In some jurisdictions, the fire code official is required to be notified before such an exercise is performed. [Ref. 405.4, 405.5, 405.6, 405.7]

TABLE 4-1 Fire and evacuation drill frequency and participation (IFC Table 405.2)

Group or Occupancy	Frequency	Participation
Group A	Quarterly	Employees
Group B[b]	Annually	All occupants
Group B[c] (Ambulatory care facilities)	Quarterly on each shift[a]	Employees
Group B[b] (clinic, outpatient)	Annually	Employees
Group E	Monthly[a]	All occupants
Group F	Annually	Employees
Group I-1	Semiannually on each shift[a]	All occupants
Group I-2	Quarterly on each shift[a]	Employees
Group I-3	Quarterly on each shift[a]	Employees
Group I-4	Monthly on each shift[a]	All occupants
Group R-1	Quarterly on each shift	Employees
Group R-2[d]	Four annually	All occupants
Group R-4	Semiannually on each shift[a]	All occupants

a. In severe climates, the fire code official shall have the authority to modify the emergency evacuation drill frequency.
b. Emergency evacuation drills are required in Group B buildings having an occupant load of 500 or more persons or more than 100 persons above or below the lowest level of exit discharge.
c. Emergency evacuation drills are required in ambulatory care facilities in accordance with Section 403.3.
d. Emergency evacuation drills in Group R-2 college and university buildings shall be in accordance with Section 403.10.2.1. Other Group R-2 occupancies shall be in accordance with Section 403.10.2.2.

EMPLOYEE TRAINING AND RESPONSE

Employees who work in buildings or occupancies that require a fire safety and evacuation plan are required by the IFC to be trained to properly respond when a fire occurs or an evacuation is required. The training must occur as part of the new-hire orientation and then at least annually thereafter. The IFC-required employee training and response procedures must address

- Recognition of fire hazards and procedures to prevent fires
- Recognition of the fire alarm and evacuation signals
- Assigned duties in the event of an emergency
- The plan for notifying, relocating or evacuating employees and occupants, and assembly points for evacuees
- A review of the site plan illustrating fire department access and fire hydrant locations
- A review of the floor plans, including primary and secondary egress paths, areas of refuge, location of portable fire extinguishers, manual fire alarm boxes and hose stations
- A review of procedures for emergency lock down and defend-in-place strategies. [Ref. 406.1, 406.2, 406.3]

In addition to training on implementing the fire safety and evacuation plan, some businesses will require employees to be trained in the proper use and operation of portable fire extinguishers, or businesses may maintain their own structural fire brigade. If personnel are trained to use portable fire extinguishers for incipient fire fighting, the IFC requires they be properly trained and know the locations of fire-fighting equipment and required personal protective clothing (Figure 4-4). [Ref. 406.3.3]

FIGURE 4-4 Plant employees training to correctly use a portable fire extinguisher to extinguish an incipient flammable liquid fire *(Courtesy of Tyco/Ansul Inc., Marinette, WI)*

Facilities storing or using hazardous materials must provide Safety Data Sheets (SDS) for each of the hazardous materials on-site. The Safety Data Sheets, formerly referred to as Material Safety Data Sheets (MSDS), must be available for review by the fire code official and for fire-fighting operations. Some jurisdictions allow SDS to be stored and available via the internet. This should only be allowed where the inspection personnel and responding fire crews have the equipment to retrieve this information as needed. New employees will also need training regarding the hazards and proper handling of the hazardous materials. [Ref. 407.2, 407.4]

PART III

Site and Building Services

Chapter 5: Fire Service Features

Chapter 6: Building Systems

Chapter 7: Interior Finish and Decorative Materials

CHAPTER 5
Fire Service Features

Fire service features include access to and into a facility, such as roadways for fire department access, a water supply for manual fire-fighting operations, a means of identifying the building through its address or other markings, and, in certain cases, a means of entering the building through the use of a key or access device under the exclusive control of the fire department; and fire-fighting operations within the building such as radio communications, fire command centers and identification of fire protection systems and utilities.

FIRE APPARATUS ACCESS ROADS

A fire apparatus access road provides vehicle access from a fire station to a facility, building or location. It is a general term inclusive of all other terms such as fire lane, public street, private street, access roadway and the drive lanes in a parking lot. The IFC provisions specify when fire apparatus access roads are required, their design and construction, required markings and requirement for barricades or gates that cross a fire apparatus access road. The IFC access road requirements are normally applied to development on private property—roads used by the public are constructed to specifications developed by the jurisdiction's Public Works or Engineering department. Design criteria that can be adopted by a jurisdiction are available in Appendix D.

An approved fire apparatus road is required for any facility, building or portion of a building constructed or moved into the jurisdiction. The road is located so that it extends to within 150 feet of all portions of the facility and the exterior walls of the first story of the building as measured by an approved route (Figure 5-1). The key term in applying this requirement is the phrase "approved route." An approved route is determined after considering the site topography and the geometry of the building in relation to the fire department access road. The measurement typically commences at the location where the fire apparatus will park on arrival and is measured along a route that can be safely used by fire fighters to a point 150 feet from an engine company. The value of 150 feet is based on a typical 1½-, 1¾- or 2-inch attack hose line carried in a horizontal cross lay, which is a conventional method of loading attack hoses on fire apparatus. The requirement is met when the entire perimeter is within the maximum distance of 150 feet. The key to properly measuring the "approved route" is to measure the distance in the same manner the attack hose will be used by a fire fighter rather than using a straight-line measurement. This would include navigating any obstacles such as perimeter fencing and gates. [Ref. 503.1.1]

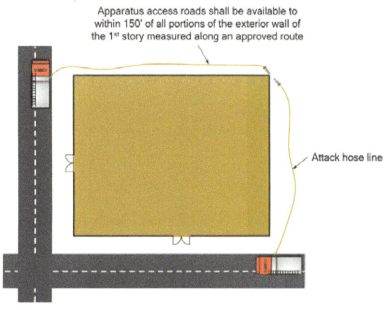

FIGURE 5-1 Measurement of the approved attack hose travel distance from a fire-apparatus access road to all portions of a building.

> **You Should Know**
>
> The key to properly measuring the "approved route," is to measure the distance in the same manner the attack hose will be used by a fire fighter rather than using a straight-line measurement. This would include navigating any obstacles such as perimeter fencing and gates. •

The IFC allows the distance from a fire apparatus access road to a building to be increased when it is protected with an approved automatic sprinkler system that complies with NFPA 13, Standard for the Installation of Sprinkler Systems, NFPA 13R, Standard for the Sprinkler Systems in Low-rise Residential Occupancies or NFPA 13D, Standard for the Installation of Sprinkler Systems in One- and Two-Family Dwellings and Manufactured Homes. The amount of the increase is not specified in the IFC; that decision rests with the fire code official. In instances where fire apparatus access roads cannot be constructed because of topography, waterways or nonnegotiable grades, the fire code official can allow the use of an alternative fire protection design in lieu of required access roads. Code officials commonly accept a building protected throughout by an NFPA 13, 13R or 13D automatic sprinkler system. The distance to access roads is not limited to a length of 150 feet where there are no more than two Group R-3 or U occupancies. Fire apparatus access roads can also be modified, or even exempted, at solar photovoltaic generation facilities. [Ref. 503.1.1]

Fire apparatus access roads must be constructed to the requirements of the jurisdiction and the IFC. The code requires fire apparatus access roads have a minimum unobstructed width of 20 feet and an unobstructed clearance of not less than 13 feet 6 inches. Road surfaces must be designed to support the imposed load of the fire apparatus and constructed of materials that are resistant to any weather conditions. The maximum permitted grade of the roadway as well as the inside and outside road turning radii are left up to the fire code official and typically based on the fire department's apparatus. The IFC limits dead-end roads to a maximum length of 150 feet before a turnaround is required (Figure 5-2). [Ref. 503.2]

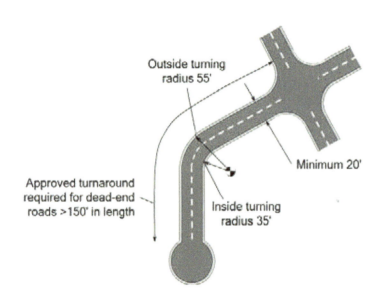

FIGURE 5-2 Turnarounds are required where dead-end roads exceed 150 feet.

Appendix D

Many of the access road requirements in IFC Chapter 5 specify that the access meet the requirements of the code official rather than provide specific dimensions. Appendix D in the IFC can be adopted by the jurisdiction and utilized to provide the specific design criteria. For example, Section 503.2.7 states that the maximum grade of the access road shall be within the limits established by the fire code official. In Appendix D, Section D103.2 specifies that the grade of an access road shall not exceed 10 percent. Appendix D is designed to provide the specific design criteria for the general requirements in Section 503. Each jurisdiction can adopt Appendix D and amend the specific numbers to fit their local needs. Table 5-1 compares the requirements in Chapter 5 and Appendix D.

TABLE 5-1 Access requirements in Chapter 5 and Appendix D

Requirement	Chapter 5	Appendix D
Minimum road width	20 feet	20 feet, and 26 feet for aerial apparatus access
Maximum grade	As required by the fire code official	10 percent
Minimum turning radius	As required by the fire code official	As required by the fire code official
Turn-around design	Must be approved	Specific design criteria
Angle of approach/departure	As required by the fire code official	As required by the fire code official
Road surface	All weather	Asphalt, concrete or other approved surface
Road design	Support the load of fire apparatus	Designed for vehicles with a weight of 75,000 pounds
Aerial apparatus access	Not specified	Required for buildings over 30 feet in height
Fire lane signs	Must be approved	Minimum size, reflective background, specific design
Multiple access routes into subdivisions	When required by the code official	More than 30 units unless sprinklered

The fire code prohibits obstructing the road with vehicles or stationary objects such as dumpsters. Fire apparatus access roads require markings using approved signs that demarcate its purpose and prohibit any obstructions (Figure 5-3). The design and placement of signs are generally governed by a state's traffic code, which provides for consistency in the sign's appearance and is also legally admissible in court if the jurisdiction issues citations for traffic violations. [Ref. 503.3, 503.4]

Security gates across fire department access roadways can slow the emergency response. An approved gate installation requires an alternate means to operate the gate during an emergency, such as a key box or a manual releasing mechanism. Gates across access roads require the approval of the fire code official. [Ref. 503.5, 503.6]

FIGURE 5-3 A fire lane must be identified to demarcate that its primary function is for fire apparatus access.

The IFC requires that all gates comply with the requirements of ASTM Standard F2200-13, Standard Specification for Automated Vehicular Gate Construction. ASTM F2200 establishes general requirements for all automated vehicle gates and construction requirements for five basic designs commonly found in the United States: horizontal slide, horizontal swing, vertical lift, vertical pivot and overhead pivot. The ASTM standard also divides gates into dif-

> **Code Essentials**
>
> A fire apparatus access road is required for most buildings to facilitate the fire department's response. The road must meet the IFC width and construction requirements. The distance from a building to a fire department access road can be increased in buildings that are protected by an approved automatic sprinkler system. •

ferent applications based on the functions of the building the gate is serving:

- Class I—One- and two-family dwellings, up to four dwellings.
- Class II—Commercial uses where general public access is desired; an example is an apartment community or a subdivision of one- and two- family dwellings.
- Class III—Industrial uses not intended to serve the public; an example is a manufacturing plant where public access is limited.
- Class IV—High-security locations, including those with guard service or security surveillance

The class of gate installed across a fire department access road is not specified by the IFC. The gate classification is dictated by the installing contractor and is selected to ensure that its design, construction and installation comply with the requirements in ASTM F2200 and the IFC. The IFC requires that security gates be equipped with a means to operate the gate during an emergency (Figure 5-4).

ASTM F2200 was developed to ensure that persons near an automatic gate are not exposed to a potential entrapment or entanglement hazard. This includes the location of the automatic gate in relation to a fixed object such as a pipe bollard or building; in such situations, the operation of a gate could create a potential crush injury to a person between the gate and the fixed object. Pedestrian gates are outside of the scope of ASTM F2200, and automated vehicular gates should not be used as pedestrian gates for the same entrapment reasons. **[Ref. 503.6]**

FIGURE 5-4 This controller for a security gate installed across a fire department access road is provided with the ability to utilize a Knox key to operate the gate.

ACCESS TO BUILDINGS

Ensuring that fire fighters can quickly access the facility completes one step in the response. To complete the response, access must be gained into the building. The IFC requires that all required exterior doors and openings be maintained accessible for use by emergency responders. This would include exit doors and windows. **[Ref. 504.1]**

The owner or occupants will have concerns over the safety and security of the individuals who use and occupy the building and its contents. Sometimes the level of security can conflict with the needs of emergency responders to access the structure and its occupants. One method that is generally regarded as a reasonable means of ensuring that the building's security is maintained while allowing for rapid fire-fighter access is the installation of a key box (Figure 5-5). The key box can only be opened by a master key that is carried by fire department officers, or for a higher level of security, can

> **Code Essentials**
>
> The IFC addresses several components for building access:
> - Building address
> - Required exterior doors and windows must be available for use
> - Key box
> - Stairway to the roof in buildings ≥4 stories. •

only be accessed when permission is electronically granted by the fire department's communication center. Many of the key box manufacturers have their equipment evaluated by a nationally recognized testing laboratory to demonstrate that it is resistant to burglars. The location of the key box, the required number of keys and the manufacturer of the key box must be approved by the fire code official. [Ref. 506.1]

The IBC requires that one of the building stairways has a means of accessing the roof when the building height is four or more stories above the grade plane. Roof access is not required when the roof is pitched and the slope is greater than 4 units vertical in 12 units horizontal, which equals an 18.3-degree slope. Roof access is provided because at a height of four stories, the roof is typically beyond the reach of ground ladders. Access to the roof can be used as a location to deploy fire streams to protect the structure from an exposure building fire and to fight the fire when rooftop equipment is involved. Even though at four stories two stairways are required, only one stairway is required to provide access to the roof. When access is provided, it can be through a penthouse or through a roof hatch. The stairway must be marked to indicate that it has roof access (Figure 5-6). [Ref. 504.3]

Certain equipment and devices in the building need to be identified for use or operation during fire fighting. Rooms containing the fire sprinkler system riser, fire alarm control panel and smoke control system panel must be identified for ready access in the event of an emergency (Figure 5-7). Typical fire department operations will also consist of shutting off the electric and fuel sources in a building involved in fire. The fire code official can require that electric meters, gas shutoff valves and solar photovoltaic switches are identified so these items can be located and turned off. [Ref. 509.1, 509.1.1, 605.11]

FIGURE 5-5 The key box contains keys for the fire department to gain access into the building.

FIGURE 5-6 The doorway is identified to indicate it serves a stairway with roof access. Note that the exterior doorway is accessible by using the fire department key box.

FIGURE 5-7 This fire sprinkler riser room is identified and also contains the controls for the smoke evacuation system.

HAZARDS TO FIRE FIGHTERS DURING ACCESS

While manual fire fighting is an inherently hazardous activity, it is safely performed daily because of the required certifications and training, assessment and implementation of fire ground risk management, personal protective equipment and the incident command system used to manage the emergency. Even with these controls and procedures, certain building features such as open building shafts may be present which create unanticipated hazards. In some cases where security concerns are significant, a building may be equipped with a device intended to incapacitate or injure a burglar. However, in the event of a fire, such a device could accidentally cause the injury or death of fire fighters. The IFC has specific requirements for protecting fire fighters from certain hazards that may be found in buildings. [Ref. 316.3]

Building shaftways are provided for elevators, material-handling equipment or for utilities such as electrical conductors or air ducts. Shaftways present a fall hazard to fire fighters because they generally are constructed to access every floor of a building. An accidental fall in a shaftway can easily injure or kill a fire fighter. To prevent the potential of an accidental fall in a shaftway, the IFC requires warning signs to identify shaft openings. If the shaft is accessible from the building exterior, a sign is required at a location that indicates the shaft's location (Figure 5-8). [Ref. 316.2.1]

FIGURE 5-8 A shaftway that is accessible from the exterior is required to be marked to indicate its location to fire fighters.

FIRE PROTECTION WATER SUPPLIES

Any new building or facility in a jurisdiction that has adopted the IFC requires a fire protection water supply capable of delivering the required fire-flow for manual fire-fighting operations. The source of the water supply can be a public water distribution system, an underground well supplied from a fire pump, a water storage tank, a reservoir or private fire service mains connected to a public water system. [Ref. 507.1, 507.2]

When a municipal or private water supply system is the source for fire-flow, a flow test is performed to demonstrate that the system is capable of providing the needed water supply. A flow test uses a test fire hydrant and a flow fire hydrant to measure the static pressure and residual pressure of the water supply system as well as the available flow rate, expressed in gallons per minute (GPM) (Figure 5-9). Static pressure is the available pressure of the water supply system with the water at rest. The static pressure can be developed by water pumps or the pressure of water in elevated tanks. Residual pressure is the available pressure when water is discharged from the flowing fire hydrant. The residual pressure is measured at the same time water is flowing from the flow hydrant. These three values are used to calculate the available fire flow. The IFC requires that fire-flow be calculated at a residual pressure of 20 PSI, which is the lowest residual pressure allowed by many state health departments and water regulatory authorities. Pressures below this value present the possibility of a water main collapse or worse, backflow of untreated sewage or wastewater into the potable water supply. [Ref. 507.4]

FIGURE 5-9 Fire hydrants are tested to determine the fire-flow provided at that fire hydrant.

Appendix B

Fire-flow is defined in IFC Appendix B as the flow rate of a water supply, measured at 20 pounds per square inch (PSI) residual pressure, that is available for fire fighting (Figure 5-10). The method used to establish the required fire-flow is determined by the fire code official. Many jurisdictions chose to adopt IFC Appendix B because it specifies the required fire-flow based on the building's height, area and construction type. The jurisdiction may use the fire-flow methods developed by the National Fire Academy or Iowa State University. In rural jurisdictions where a conventional water supply system is unavailable, Appendix B allows the fire code official to apply water supply requirements in the *International Wildland-Urban Interface Code* or NFPA 1142, *Standard on Water Supplies for Suburban and Rural Fire Fighting*. These two documents provide alternative fire-flow requirements based on the use of natural or man-made water sources, or the use of mobile water tankers as the mechanism for water delivery to the site. The specific method of determining the required water supply should be established in the adopting ordinance or in a written policy to provide for consistency in application. [Ref. 507.3, B103.3]

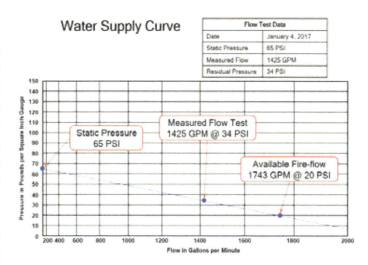

FIGURE 5-10 Plotting the flow test results on a water supply curve can determine the available water supply at 20 PSI.

Code Essentials

The static pressure, residual pressure and flow rate (measured in gallons per minute) are required to determine the fire-flow. Fire-flow is calculated at a residual pressure of 20 PSIG.

Inspection and maintenance

Fire protection water supplies are commonly supplied from private hydrants with private fire service mains constructed on private property to supply the required fire-flow. The IFC requires the construction of private fire protection water mains be in accordance with NFPA 24, *Standard for the Installation of Private Fire Service Water Mains and Their Appurtenances*. All portions of the exterior wall of a building must be located within 400 feet of a fire hydrant, or approved water supply, on a fire apparatus access road. The distance is measured along an approved route using the same method prescribed by the fire code for locating buildings in relation to fire apparatus access roads. The distance is measured using a path that fire apparatus will unload water supply hoses on the roadway and the path that hoselines will be extended to the building, instead of a straight line measurement from the fire hydrant to the building (Figure 5-11). When a building is protected by an automatic

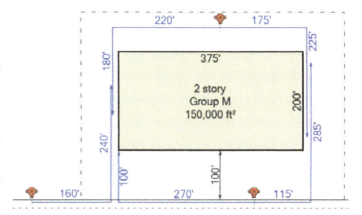

FIGURE 5-11 The water supply or fire hydrants must be within 400 feet of all portions of the exterior wall of the first floor of the building.

FIGURE 5-12 Flushing of a private fire hydrant is intended to remove and discharge any sediment or debris in the private fire-protection water main.

FIGURE 5-13 This fire hydrant would be difficult to operate with the wall obstructing access to the valves.

sprinkler system installed in accordance with NFPA 13 or NFPA 13R, the fire code official can increase the distance requirement to 600 feet. For Group R-3 and U occupancies, the travel distance is also 600 feet. [Ref. 507.5.1]

A fire protection water supply system requires annual maintenance to ensure the system will deliver the required fire-flow (Figure 5-12). The IFC refers to the procedures in NFPA 25, Standard for the Inspection, Testing and Maintenance of Water-Based Fire Protection Systems for inspection and testing of the fire water supply. Critical components of the fire protection water supply must be inspected and tested, and records maintained for fire code official review, as follows:

- Private fire hydrants: inspected annually and after each use; flow test annually
- Private fire protection water mains: inspect exposed piping annually; flow test every 5 years
- Strainers installed in private fire protection water mains: inspection and maintenance after each use but at least annually. [Ref. 507.5.3]

To ensure that fire hydrants are always accessible, the IFC prohibits their obstruction and requires a minimum 3-foot clearance around their circumference (Figure 5-13). When a fire hydrant is located in an area subject to impact by motor vehicles, a means of vehicle impact protection complying with the IFC is required (Figure 5-14). [Ref. 507.5.5, 507.5.6]

FIGURE 5-14 The guard posts around the fire hydrant provide vehicle impact protection.

EMERGENCY RESPONDER RADIO COVERAGE

Public safety radio systems in numerous regions of the United States have been converted from or are in the process of being converted from analog signaling systems to digital signaling systems. Because they are digital systems, the portable radios used by fire fighters and emergency responders require less power for transmitting audio or electronic signals compared with that required for the former analog systems (Figure 5-15). These systems also offer greater functionality, because each radio is assigned its own digital address within the public safety radio system, based on the user's roles and responsibilities at emergency incidents. These systems provide for interoperability, meaning a fire fighter can now directly communicate with police officers, emergency medical personnel and any other person assigned to a particular emergency incident. This interoperability can be extended to other emergency responders in one or more regions of a state to one or more states or areas in the United States.

The IFC emergency responder radio coverage provisions are concerned with the reliability and usability of portable radios when used inside buildings. Portable radios are used to communicate with other emergency responders, the incident commander or the public safety communications center. Building construction features and materials can absorb or block the radio frequency energy used to carry the signals inside or outside of the building. Blockage or absorption of the radio frequency signal can prevent a critical message sent by an emergency responder from being received and acknowledged. Depending on the incident, this loss of information can place emergency responders in greater danger or may prevent an injured or disoriented emergency responder from communicating for assistance.

FIGURE 5-15 A fire fighter using a portable radio inside a building.

Section 510 addresses requirements for in-building coverage of emergency responder radios. The requirements in this section apply to both analog and digital radio systems and are applicable to all buildings. Section 510.1 requires that approved radio coverage for emergency responders is available within all new buildings. Approved radio coverage is based on the ability of the public safety communication system to receive and send an acceptable signal while inside the building. The concept is to provide a level of communication close to the level of communication outside of the building at the street. However, if radio communications are poor outside the building, the building owner is not responsible to correct that situation. [Ref. 510.1]

The size and construction of the building and the number of floors or basement levels can affect the radio signals as they pass through the building walls. Radio coverage is considered acceptable if the delivered audio quality (DAQ) is 3.0 or better (see sidebar). This minimum DAQ must be maintained in 95 percent of all areas on each floor of a building. If the building does not interfere with the radio signals, then no equipment or modifications are necessary in the building. If the signal clarity is inadequate, then amplifiers, repeaters or other radio equipment must be installed to strengthen the signal within the building. [Ref. 510.4]

Discussions with public safety radio professionals found that based on current radio technologies, some method to provide emergency responder radio coverage is typically necessary in any building with one or more basements or below-grade levels, any underground building or any building more than five stories in height. In most wood frame or mixed construction Group R-1 and R-2 occupancies, single-family dwellings, townhouses and buildings with an area less than 50,000 square feet without basements, there is little concern for loss of radio signal strength inside the building or inability to transmit to an outdoor receiver.

Exception 1 of Section 510.1 allows the code official to accept the installation of a wired communication system in lieu of the radio coverage. When this exception is applied, the concurrent approval of the fire and building code officials is required. The wired communication system does not provide communication throughout 95 percent of the building, but provides communication in specified locations. It would be designed to operate between a specified location, such as a fire command center, and elevators, elevator lobbies, emergency and standby power rooms, fire pump rooms, areas of refuge and at each floor level within enclosed exit stairways. [Ref. 510.1]

Exception 2 of Section 510.1 allows the code official to waive the requirement when it is determined that emergency responder radio coverage is not needed. Exception 2 does not give any criteria as to when these requirements are applicable or which buildings can be exempted.

Exception 3 of Section 510.1 addresses situations where the constant operation of the radio equipment has a negative impact on the normal operations of that facility. In these cases, the code official can allow for the emergency responder radio coverage system to be in a normally "OFF," or non-transmitting, mode. The system would be automatically activated when needed, although the code does not specify the method of automatic activation. This could be upon water flow from the fire sprinkler system, activation of the fire alarm, or any other trigger approved by the code official.

The IFC also specifies the requirements for emergency responder radio coverage in existing buildings. As previously noted, when radio coverage is not adequate, Section 510.1 Exception 1 permits the installation of a wired communication system. However, if a building contains an existing wired communication system, and the system cannot be repaired or is undergoing replacement, Section 1103.2 allows the code official to require the installation of a system to provide emergency responder radio coverage. If an existing building is known to have inadequate radio coverage Section 1103.2 allows the code official to establish a time frame for compliance with Section 510.1. This could be either a wired communication system or a building radio amplification system. Prior to construction, there are times when it will be difficult to ensure that radio coverage will not be affected by the building itself. This is the reason this requirement is also applicable to existing facilities. [Ref. 510.2]

> **Code Essentials**
>
> Delivered Audio Quality (DAQ)
>
> - **DAQ 1:** Transmission is unusable. Speech is present but unreadable.
> - **DAQ 2:** Speech is understandable with considerable effort. Frequent repetition is needed due to noise or distortion.
> - **DAQ 3:** Speech is understandable with slight effort. Occasional repetition is needed due to noise or distortion.
> - **DAQ 3.5:** Speech is understandable with slight effort. Occasional repetition is needed due to noise or distortion.
> - **DAQ 4:** Speech is easily understood. There is occasional noise or distortion.
> - **DAQ 4.5:** Speech is easily understood. There is infrequent noise or distortion.
> - **DAQ 5:** Speech is easily understood.

Determining whether a building has adequate or inadequate radio coverage is most easily accomplished by having qualified personnel perform functional radio tests inside areas of the existing building. For buildings under construction, design professionals may issue a letter of intent of compliance to the code official that documents that the owner will comply with the requirements and install either a wired communication system or a signal booster system if the functional radio test fails to transmit a signal outside the building or to other radios inside the building after the construction is substantially completed. Some jurisdictions are requiring a third-party review of the building features before issuing a building permit, but this can be difficult because the radio frequency signal absorption and deflection characteristics of different building materials are not completely understood by the engineering community. If code officials wish to perform a functional test, Section 510.6.4 authorizes the code official to enter a building to perform the tests at any reasonable time.

In new buildings, the fire code official can apply the IFC acceptance testing criteria as a means of confirming that portable radios will provide adequate emergency responder radio coverage. Persons performing these tests should have a valid Federal Communication Commission General Radio Operators License and a certification as to their understanding of the in-building amplification equipment as specified in Section 510.5.2. Code officials may not be completely confident in performing such tests and should consider applying the requirements in Section 107.2 to treat this test as a special inspection. [Ref. 510.5.3]

One of the normal fire-fighting operations is to shut down the electrical power to a building during the fire emergency. If this occurs in a building with powered equipment providing emergency responder radio coverage, the radio communication ability would be lost. To eliminate this possibility, the emergency responder radio coverage system must be provided with standby power which will operate the system at 100-percent capacity for 12 hours. [Ref. 510.4.2.3]

> **You Should Know**
>
> A building constructed in a jurisdiction that has adopted the IFC will require fire apparatus access roads, a water supply capable of supplying the required fire-flow and a means of access into the building. Private fire hydrants and fire protection water mains may be required as a source of fire-flow. All buildings require emergency responder radio coverage. ●

CHAPTER 6
Building Systems

Building systems include fuel-fired appliances, standby and emergency power, elevator operation by emergency responders, mechanical refrigeration systems and cooking operations. The provisions in IFC Chapters 6 and 12 set forth minimum requirements for these building systems. Some of the building systems use hazardous materials, such as mechanical refrigeration, fuel-fired appliances and commercial cooking operations. The location of these requirements in IFC Chapter 6 results in mitigating hazards specific to the operation, rather than regulation under the more general hazardous materials requirements in IFC Chapter 50 (see Part VI of this book). These systems are not exempted from regulations; rather, they are exempted from regulation under Chapter 50, because they comply with other requirements in the code its adopted standards and the requirements in the *International Mechanical Code* (IMC) and *International Building Code* (IBC).

FUEL-FIRED APPLIANCES

The IFC contains requirements for the installation and operation of fixed fuel-fired appliances. An appliance is an apparatus or device using fuel gas or fuel oil to produce light, heat, power, refrigeration or air conditioning. The IFC requires that all gas-fueled appliances be installed in accordance with the requirements of the *International Fuel Gas Code* (IFGC), liquid-fueled appliances must comply with the IMC. [Ref. 603.1, 603.1.2]

In the U.S., the two fuel gases commonly consumed are natural gas and liquefied petroleum gas (LP-gas). Natural gas is a mixture of flammable gases whose primary constituent is methane. LP-gas is a mixture of ethane, methane, propane and butane. Natural gas and LP-gas are colorless and odorless flammable gases. The density of natural gas makes it lighter-than-air when it is released, while LP-gas is heavier-than-air. The U.S. Department of Transportation requires both gases be odorized when they are shipped using a pipeline. An odorant like methyl mercaptan is added so any leak is detectable by a distinctive odor when the volume of the release is at 25 percent of its lower flammable limit.

Fuel oil is defined in the IMC as a hydrocarbon distillate with a flash point temperature of 100°F or more, which classifies the material as a combustible liquid based on the classification criteria in IFC Chapter 57. (See Chapter 18 in this book for an explanation about classification of flammable and combustible liquids.) Fuel oils include kerosene, diesel fuel and biodiesel.

All fuel-fired equipment is required to be installed in accordance with the manufacturer's instructions and modification of the equipment must be performed in accordance with the requirements of the original equipment manufacturer. Adequate access is required around equipment so it can be maintained (Figure 6-1). The IFGC and IMC contain extensive requirements for equipment access. [Ref. 603.1.2, 603.1.5]

Fuel oil is used as a source of heat and to operate engine-driven fire pumps or generators. Petroleum products used in fuel-fired equipment must meet the original equipment manufacturer's specifications, and the use of a petroleum product that is contaminated with gasoline is prohibited by the IFC. In Group F, M and S occupancies, the business may choose to use heaters that consume waste oil. Waste oil heaters and boilers are commonly found in repair garages and manufacturing plants using petroleum formulated cutting fluids (Figure 6-2). Waste oil fuel-fired appliances are required

> **Code Essentials**
>
> Liquefied petroleum gas (LP-gas) and methane (natural gas) are two forms of fuel gas. All fuel gases, other than hydrogen, are odorized so leaks can be easily detected. Kerosene and No. 2 diesel fuel are examples of fuel oils. The IMC specifies that fuel oils are kerosene or any hydrocarbon oil with a flash point temperature of 100°F or more.

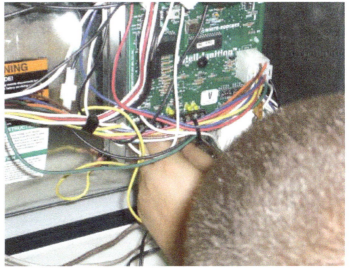

FIGURE 6-1 Fuel-fired equipment must be installed with adequate access to facilitate maintenance.

by the IFC to be listed as heat-recovery devices. These heaters are limited to the consumption of used crankcase oil up to Society of Automotive Engineers 50 weight, used transmission and hydraulic fluids, and No. 2, 3 and 4 fuel oils. If waste oil heaters are installed in a repair garage, they also must comply with the requirements in IFC Chapter 23. [Ref. 603.1.4]

The amount of fuel oil stored is dependent on the type of storage tank and the level of fire protection. When stored above ground, the IFC permits the installation of the storage tank inside or outside a building. Fuel oil storage outside of buildings is limited to a volume of 660 gallons. If more than 660 gallons of fuel oil is required, the design of the storage tank and fuel oil delivery system must comply with the requirements of NFPA 31, Standard for the Installation of Oil-Burning Equipment. NFPA 31 defers to the requirements of NFPA 30, *Flammable and Combustible Liquids Code* for tanks storing more than 660 gallons of fuel oil. [Ref. 603.3.1]

FIGURE 6-2 Waste-oil burners *(Courtesy of Clean Burn Energy Systems, Leola, PA)*

Requirements for the indoor storage of fuel oil are based on the storage amount, type of storage tank and the fire protection provided. Storage of up to 660 gallons of fuel oil is allowed provided the storage tanks comply with UL 80, Standard for Steel Tanks for Oil-Burner Fuels and Other Combustible Liquids, UL 142, Standard for Steel Aboveground Tanks for Flammable and Combustible Liquids or UL 2085, Standard for Protected Aboveground Tanks for Flammable and Combustible Liquids. Quantities from 660 gallons to 1,320 gallons can be stored in tanks listed to UL 142 or UL 2085 provided the building is provided with an automatic sprinkler system. Fuel oil storage is allowed in up to 3,000 gallons when the fuel oil is stored in a tank listed to UL 2085 and the room is provided with an automatic sprinkler system (Figure 6-3). [Ref. 603.3.1]

FIGURE 6-3 PASTs for fuel oil located in sprinklered buildings can have a capacity up to 3,000-gallons.

The requirements for above-ground fuel oil storage inside buildings provide for several levels of protection and quantity needs. The quantity of fuel oil needed to operate an emergency or standby generator is increasing as electrical demands increase along with desires to maintain a constant supply of power. Rooms with fuel oil tanks supplying internal combustion engines, such as generators, must be separated by 1-hour fire-resistance-rated construction from the remainder of the building. The 1-hour separation is not required when a protected above-ground storage tank (PAST) is used as the storage tank. There is a three-tier approach to protecting fuel oil storage tanks inside a building. Table 6-1 contains the allowed quantities based on the type of storage tank and the level of fire protection provided.

TABLE 6-1 Maximum capacity of fuel oil based on type of tank and fire protection provided

Tank design	Nonsprinklered building	Automatic fire sprinklers		1-HR room construction
		Sprinklers in the room	Sprinklers throughout the building	
UL 80	660 gallons	660 gallons	660 gallons	Required
UL 142	660 gallons	660 gallons	1,320 gallons	Required
UL 2085	660 gallons	3,000 gallons	3,000 gallons	Not Required

A PAST is a tank listed to UL 2085. PASTs are shop-fabricated above-ground storage tanks that have been subjected to a fire test that replicates exposure to a 2-hour flammable liquid pool fire. PASTs are constructed with integral secondary containment and are evaluated for vehicle impact and bullet resistance. All openings on a PAST are located at the top of the storage tank, which reduces the potential for liquid leaks; openings below the liquid level in PASTs are prohibited by the IFC. [Ref. 603.3.2.1]

To limit the potential of a fire involving the PAST, the entire room housing the storage tank must be protected by an approved automatic sprinkler system that complies with the requirements in NFPA 13. The function of the PAST is limited to the supply of fuel oil to fuel-fired appliances, including generators—it cannot be used for any other purpose. The PAST cannot be located more than two stories below the building's grade plane. Fuel oil piping must be liquid tight and comply with the requirements of the IMC (Figure 6-4). [Ref. 603.3.2.1, 603.3.2.4]

Fuel-fired appliances are designed to operate as either vented or unvented appliances. They are commonly used to heat a room or an area of a room. A vented appliance directly exhausts combustion gases produced by the fuel gas or fuel oil outside of the building. Vented devices commonly use air from the room in which they are located as their source of supply air.

Unvented heaters also use indoor air as their

Code Essentials

The amount of fuel oil allowed inside a building is limited by the IFC, based on the type of storage tank. Above ground and inside of buildings, a maximum of 3,000 gallons of fuel oil can be stored in a PAST. •

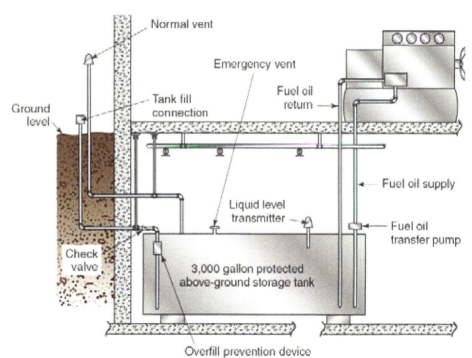

FIGURE 6-4 This 1,000-gallon protected above-ground storage tank is located inside a building and supplies an engine-driven generator.

FIGURE 6-5 This portable LP-gas heater is an unvented fuel-fired appliance.

> **Code Essentials**
>
> Fuel-fired appliances can be vented or unvented. The IFC prohibits the installation of unvented heaters in most occupancies because of the potential for carbon monoxide poisoning. •

FIGURE 6-6 Portable outdoor gas-fired heating appliances are allowed when in compliance with Section 603.4.2.1.1.

source of combustion air. However, these heaters release the combustion byproducts into the room. Most portable unvented heaters use kerosene or LP-gas as fuel (Figure 6-5). When these heaters are properly operated and maintained, these systems can be up to 98-percent efficient. Although even with this high efficiency, the constant use of unvented heaters inside a building can introduce carbon monoxide into a room if fresh makeup air is not introduced. Carbon monoxide molecules have a higher affinity to attach to red blood cells than oxygen. If an atmosphere contains enough carbon monoxide in the air, the carbon monoxide can displace the oxygen, preventing it from being carried into the body. Carbon monoxide poisoning can cause injury or death.

Because of the concerns over carbon monoxide poisoning, as well as issues of fire safety since these heaters emit radiant and convective energy, the IFC has specific provisions concerning portable unvented heaters. Unvented heaters are prohibited in assembly, educational, institutional and residential occupancies. Listed and approved heaters are allowed in one- and two-family dwellings.

Although listed devices are allowed in one- and two-family dwellings, their operation and location is still regulated. Portable unvented heaters are not permitted to be located in or to obtain combustion air from sleeping rooms, bathrooms or closets. **[Ref. 603.4.1]**

Another form of unvented heaters is portable outdoor gas-fired appliances. These heaters are especially popular at outdoor seating areas of restaurants and similar occupancies. Being unvented appliances, they are not designed or listed for indoor use. The IFC prohibits the use of portable outdoor gas-fired heating appliances inside buildings, tents, canopies and membrane structures (Figure 6-6). Use of these heaters is also prohibited on exterior balconies of apartment buildings and similar multifamily residential areas, as prescribed in NFPA 58, *Liquefied Petroleum Gas Code*. NFPA 58 prohibits the use or storage of liquefied petroleum-gas (LP-gas) containers on exterior balconies of apartments when the container volume is greater than 1.08 pounds of propane. **[Ref. 603.4.2.1.1]**

Portable outdoor gas-fired heating appliances are required to be listed and approved by the fire code official (Figure 6-7). For patio heaters, the applicable standard is American National Standards Institute (ANSI) Z83.26, Standard for Gas-Fired Outdoor Infrared Patio Heaters. One safety feature required by the ANSI standard is the connection between the LP-gas cylinder and the hose supplying the appliance's burner. The standard requires that the hose connected to the appliance be equipped with a Compressed Gas Association (CGA)

790 fitting. A CGA 790 fitting provides three separate safety features. First, the fitting has a thermal link that is designed to activate at temperatures of 200° to 250°F that stops the flow of LP-gas in the event of a fire. Second, the fitting requires a positive connection to the cylinder before LP-gas can flow into the appliance. Third, the fitting is equipped with an internal excess flow control valve. An excess flow control valve is designed to stop the flow of a gas or liquid in the event of hose or pipe rupture. [Ref. 603.4.2.2.1]

When portable gas-fired heating appliances are utilized, specific safeguards are included to minimize ignition of adjacent materials and maintain a safe evacuation route for the occupants. The portable gas-fired appliance must be equipped with a tip-over switch that will automatically shut off the appliance when it is tilted more than 15 degrees. This prevents the appliance from spreading fire if it were to be accidentally knocked down. The maximum gas container size cannot exceed 20 pounds. A minimum separation of 5 feet is required between the appliance and buildings, combustible awnings or overhangs, decorations and exits, and exit discharges. Refill containers for gas-fired appliances shall not be stored inside the building unless each container is no larger than 2.7 pounds water capacity. [Ref. 603.4.2.1, 603.4.2.2, 603.4.2.3, 6109.9]

MECHANICAL REFRIGERATION

The IFC contains requirements for mechanical refrigeration systems which are used in conjunction with the requirements in the IMC. The IMC regulates all mechanical refrigeration systems and provides classification for the various refrigerants. [Ref. 605.1]

The IFC contains provisions for specific refrigerants that have hazardous characteristics, i.e., flammable refrigerants, toxic refrigerants and ammonia. These refrigerants have characteristics that the IFC is concerned about.

FIGURE 6-7 Portable outdoor gas heater *(Courtesy of Infrared Dynamics, Yorba Linda, CA)*

TABLE 6-2 IMC refrigerant classification system

Classification	Description	Examples
Toxic		
Class A (nontoxic)	Occupational exposure limit (OEL) ≥ 400 ppm	Ethane, R134a, water
Class B (toxic)	Occupational exposure limit (OEL) < 400 ppm	R123, ammonia
Flammable		
Class 1 (nonflammable)	No flame propagation	R22, CO_2, water
Class 2 (moderate flammability)	Lower flammable limit > 0.00625 lb/ft³ and Heat of combustion < 8,169 Btu/lb	R32, ammonia
Class 2L (low flammability)	Lower flammable limit > 0.00625 lb/ft³ and Heat of combustion < 8,169 Btu/lb and Burn velocity < 10 cm/sec	R1234yf, R1234ze(E)
Class 3 (high flammability)	Lower flammable limit ≥ 0.00625 lb/ft³ and Heat of combustion ≥ 8,169 Btu/lb	Ethane, propane

FIGURE 6-8 This refrigeration machinery room requires an emergency shutoff switch to shut down the compressors and pumps, and a switch to turn on the ventilation fans.

The regulations in the IFC do not cover all refrigerants. All refrigerants are classified for both toxicity and flammability. A typical refrigerant such as R12 is classified as Group A1. It is considered nontoxic and nonflammable. However, R12 is also known to be an ozone depleter and contributes to global warming, so it is being phased out of the market. Many new refrigerants are being developed as a result, which has also led to a new classification with regard to flammability—Class 2L refrigerants. These refrigerants are a subclass of Class 2, however these refrigerants require a stronger source of ignition, the lower flammable limit is higher, and the velocity, or speed, of flame spread is low. All of these characteristics result in a refrigerant that has a low probability of ignition and a low impact if ignition does occur.

Refrigerant detection is required where the refrigerant is toxic, flammable or ammonia and a machinery room is provided (Figure 6-8). Machinery rooms also require an emergency shutoff switch which will shut down the compressors and pumps, and another switch which will turn on the ventilation fans. [Ref. 605.9]

Refrigeration systems containing more than 6.6 pounds of refrigerant require an emergency pressure control system. This system is designed to detect an overpressure situation, open a pressure relief device which discharges to the next lower pressure zone in the refrigeration system and automatically shut down the system. Refrigeration systems consist of at least two pressure zones and many have multiple pressure zones. Each pressure zone will have a pressure relief device which activates as described. The lowest pressure zone has no lower pressure zone in which to discharge the excess pressure, so it will discharge through a treatment system for toxic refrigerants, through a flare for flammable refrigerants, or through an ammonia diffusion system for ammonia. There is an allowance for ammonia to discharge through a flare, a treatment system or directly to the atmosphere when approved by the code official. [Ref. 605.10, 605.12]

ELEVATORS

The only buildings that require elevators are those that have a story that is four or more stories above or below the level of exit discharge to provide an accessible means of egress for mobility-impaired persons. To apply this requirement correctly, it must be understood that the level of exit discharge is the first floor (See Figure 6-9). Four stories above the level of exit discharge results in a five-story building. [Ref. 1009.2.1]

When a building has a floor level located more than 120 feet above the lowest level of fire department access, the IBC requires a fire service access elevator. In all other buildings, elevators are provided to facilitate the movement of the building occupants, and, as such, they must be installed in accordance with the applicable IBC requirements. [Ref. 606.4]

Requirements for the design, construction and testing of elevators are contained in ASME/ANSI A17.1, *Safety Code for Elevators and Escalators*. Section 607.1 and ASME A17.1 require all new elevators to be equipped with Phase I and Phase II emergency in-car operation. In Phase I service, a smoke detector is installed in each elevator lobby. Activation of a lobby smoke detector causes the elevator car to be captured and recalled to the designated floor, which is usually the ground floor of a building. Upon arrival, the car door opens and the elevator is no longer operable by the occupants. In the event a smoke detector fails or if emergency responders wish to use the elevator such as for the transportation of equipment in the treatment of a patient, a key switch is provided in the elevator lobby (Figure 6-10). This key switch overrides the automatic function of the elevator and captures the elevator and recalls it to the floor level where the switch was activated. [Ref. 606.1]

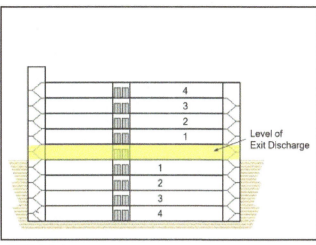

FIGURE 6-9 A building with a floor level that is four or more stories above or below the level of exit discharge must be provided with an elevator.

FIGURE 6-10 A Phase I emergency in-car operation feature

Code Essentials

Phase I service is required in
- New elevators
- Existing elevators in unsprinklered buildings with a travel distance of 55 feet above or 25 feet or more below the lowest level of fire department vehicle access.

Phase II service is required in all new elevators. Phase I service captures and returns the elevator to a designated level. Phase II service provides fire fighters with exclusive control of the elevator.

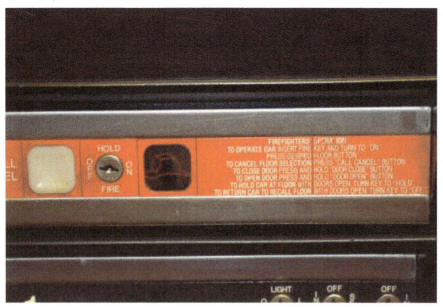

FIGURE 6-11 A Phase II emergency in-car operation feature

Phase II emergency operation allows fire fighters to control the elevator and travel to any floor in the elevator bank (Figure 6-11). The operating controls are located inside of the elevator car. When placed into the fire service mode, the elevator can only be operated by personnel in the elevator car. To return the elevator to normal service, the elevator must be reset using the Phase I switch. [Ref. 606.1]

Existing elevators are required to be retrofit with Phase I and II operations in Group I-2, Condition 2 occupancies and Group B ambulatory care facilities. In all other occupancies, Phase I and II operations are required when the elevator car travels more than 25 feet above the main floor or 25 feet below the main floor, with the following exceptions:

- Buildings with floor levels not greater than 55 feet above or 25 feet below the lowest level of fire department vehicle access and automatic closing, fire-rated doors are provided over the elevator doors activated by smoke detectors installed in the elevator lobbies.
- Buildings with floor levels not greater than 55 feet above or 25 feet below the lowest level of fire department vehicle access and the building is fully sprinklered.
- Buildings provided with at least one elevator equipped with Phase I and II operations and sprinklers throughout the building.

[Ref. 1103.3.2]

In the event that a lobby smoke detector activates, all of the elevators in the bank will be recalled to the designated floor. Elevators are generally not permitted to be used as a means of egress component by the IBC. To ensure that occupants do not attempt to use an elevator as a means of egress, a standard pictorial sign is required in each elevator lobby that instructs the occupants to use exit stairs when a building is being evacuated (Figure 6-12). [Ref. 606.3]

FIGURE 6-12 A pictorial sign is required at each elevator lobby instructing occupants to use the exit enclosure stairs during a building fire.

When elevators are provided in a building, at least one elevator car must be available to all floor levels and it must be sized to accommodate an ambulance gurney. This elevator car is to be identified for emergency use with the international symbol for emergency services, the star of life, on the hoistway door (Figure 6-13).

FIGURE 6-13 The Star of Life symbol on an elevator hoistway indicates that the elevator is large enough to accommodate a gurney.

COMMERCIAL COOKING OPERATIONS

In a 2017 report by the U.S. Fire Administration, an estimated 30,900 cooking fires occurred in nonresidential buildings, resulting in an estimated $56.5 million in property loss^2. Although this report contains no estimates of deaths, the potential for fire fatalities exists in any building or property where people congregate. Based on this loss history, commercial cooking operations present a fire- and life-safety threat that requires close supervision. Three main areas of concern are

1. Exhaust hood
2. Cooking oil storage
3. Automatic fire-extinguishing system.

Commercial kitchen exhaust hoods

Commercial cooking appliances are used in a commercial food service establishment for cooking food. Commercial cooking appliances require a local exhaust ventilation system to remove heat, vapors, steam, smoke or odors produced during food preparation, cooking or cleaning activities. The requirements for exhaust systems are set forth in the IMC, along with designation of which commercial cooking appliances require a Type I hood. A Type I hood is designed for the removal of grease-laden vapors and smoke (Figure 6-14). [Ref. 607.1]

FIGURE 6-14 A commercial cooking operation with a Type I exhaust hood.

2. *Nonresidential Building Cooking Fire Trends* (2006 - 2015), U.S. Fire Administration, May 2017.

Certain cooking appliances produce less grease than others. Exception 4 provides for cooking appliances that produce volumes of grease so minimal that they do create the need for a Type I hood. An electric conveyor pizza oven may fit this exception. The appliance must be electric powered and tested to confirm that the appliance emits no more than 5 mg/cm^3 when operated at an exhaust flow rate of 500 cfm. This is a specific test that is found in UL 710B, Standard for Recirculating Systems. The cooking appliance does not need to be listed to this standard, but it must have documented proof from a testing laboratory that it passes this specific test. Exception 2 allows for appliances that are listed to UL 710B. It should be noted that if the appliance does not require a Type I hood, a fire-extinguishing system is not required either.

Type I hoods and their exhaust ducts will collect the fats produced during the cooking processes. The amount of grease and smoke that is generated is dependent on the method of cooking, the type of protein being cooked and the intensity of the cooking operations. Accordingly, the IFC establishes prescriptive inspection frequencies in Table 607.3.3.1 (see Table 6-3).

TABLE 6-3 Commercial cooking system inspection frequency (IFC Table 607.3.3.1)

Type of cooking operation	Frequency of inspection
High-volume cooking operations such as 24-hour cooking, charbroiling or wok cooking	3 months
Low-volume cooking operations such as places of religious worship, seasonal businesses and senior centers	12 months
Cooking operations utilizing solid-fuel-burning cooking appliances	1 month
All other cooking operations	6 months

The most restrictive inspection frequency is for cooking operations using solid fuels, such as barbeque pits and meat smokers. When appliances use a solid fuel such as wood or charcoal, a minimum monthly inspection frequency is required. Cooking operations using charbroiling or woks require a minimum 3-month inspection frequency, as do high-volume cooking operations found in 24-hour restaurants. The frequency of inspection is reduced to 12 months for cooking activities involving seasonal businesses, places of worship and facilities that provide care for the elderly. All other cooking operations are subject to a 6-month inspection frequency. Depending on the nature of the commercial cooking activities, a single kitchen could have different inspection frequencies. For example, a restaurant that serves smoked meats must inspect the smoking appliances monthly, whereas other equipment would require inspection every 3 or 6 months, depending on the volume of cooking being performed. [Ref. 607.3.3.1]

If the inspection reveals an accumulation of grease, the hood, grease removal devices, exhaust fans and ducts are required to be cleaned (Figure 6-15). The required inspection will reveal whether cleaning is necessary or not. A tag must be affixed to the exhaust hood after the inspection is completed to indicate the date of the most recent inspection. When the system is cleaned, a tag will also be affixed to the hood with the date of the cleaning (Figure 6-16). [Ref. 607.3.3.2]

FIGURE 6-15 Cleaning of Type I hoods that serve commercial cooking operations is a required activity to limit the amount of available fuel that could spread fire.

FIGURE 6-16 An inspection and service tag is to be provided to document that the commercial kitchen exhaust hood was inspected or cleaned. *(Courtesy of Flue Steam Inc., Culver City, CA)*

Cooking oil storage

Cooking oils used in deep-fat fryers need to have flash points higher than the temperatures they are exposed to during cooking operations. This does not mean that they will not ignite and burn, just that the temperature needed to reach flash point is higher. As such, the IFC states that cooking oils are treated as Class IIIB liquids unless actual flash point testing proves otherwise. Both fresh oil and waste oil are classified as Class IIIB combustible liquids. [Ref. 608.1]

These oils are combustible liquids and they need to be properly stored, since quantities in a commercial kitchen can be large. Cooking oil storage methods must meet the requirements in IFC Section 608 and NFPA 30, *Flammable and Combustible Liquids Code* with regard to proper storage container and location. Requirements include the type of container, proximity to ignition sources, height of storage, etc. Cooking oil is not required to be stored in a tank; it could be stored in the DOT shipping container. The most common form of storage in restaurants is referred to as a jug-in-a-box, which is a 4.5 gallon container in a cardboard box. When cooking oil supplies are contained in a tank exceeding 60 gallons, the storage tank must be designed for the storage of combustible liquids or cooking oils. [Ref. 608.1, 608.2, 608.3]

FIGURE 6-17 Delivery and recovery of cooking oil utilizing oil storage tanks (*Courtesy of Restaurant Technologies, Inc.*)

The use of tanks for the storage of cooking oil is becoming quite common. The storage tanks eliminate the need for kitchen employees to manipulate each jug-in-a-box through the facility, manually dispense the contents into the fryer and properly discard the empties. Tubing is installed between the fryers and the tanks, enabling fresh oil to be pumped to the fryers and waste oil to return to the waste oil tank. Additional connections are installed outside the building for the delivery of fresh oil and the collection of the waste oil (Figure 6-17).

Cooking oil storage tanks can be constructed of metallic or nonmetallic materials. Whether of metallic or nonmetallic construction, the tanks must be listed. Cooking oil storage tanks of metallic construction must be listed to UL 142, Steel Aboveground Tanks for Flammable and Combustible Liquids or UL 80, Steel Tanks for Oil-Burner Fuels and Other Combustible Liquids. Nonmetallic cooking oil storage tanks must be listed for the storage of cooking oil and evaluated to handle cooking oil at the anticipated temperatures. Nonmetallic cooking oil storage tanks are limited to a capacity of 200 gallons each. Nonmetallic tanks must be listed to UL 2152, Outline of Investigation for Special Purpose Nonmetallic Containers and Tanks for Specific Combustible or Noncombustible Liquids (Figure 6-18). When storage tanks are utilized, it is quite common for two tanks to be installed, one tank for fresh cooking oil and one tank for waste cooking oil. [Ref. 608.2, 608.3]

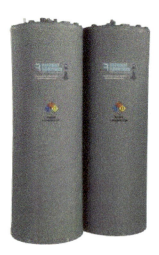

FIGURE 6-18 Nonmetallic cooking oil storage tanks listed to UL 2152 (*Courtesy of Restaurant Technologies, Inc.*)

Venting is required on flammable or combustible liquid storage tanks, and cooking oil storage tanks are no different. Since the flash point of cooking oils is so high, the normal vent is allowed to terminate inside the room where the tank is located. Emergency vents are required to prevent rupture during a fire exposure. Metallic storage tanks must be equipped with an emergency vent. Nonmetallic tanks can either be provided with an emergency vent device, or the tank construction itself can be designed to fulfill the emergency venting requirement. Emergency vents on both metallic and nonmetallic tanks are allowed to terminate inside the building. [Ref. 608.5]

Automatic fire-extinguishing systems

The fire hazard is greatest in cooking operations where grease-laden vapors are produced. The grease travels up through the exhaust hood and duct, and as it cools it accumulates along the interior lining of the hood and duct. With continued use, this grease layer thickens and the heat exhausted through the hood and duct keep this layer of grease at an elevated temperature. When a fire occurs at the cooking surface and flames extend up into the hood, the heated grease is easily ignited and fire can quickly extend throughout the entire duct.

Section 507.1 of the IMC requires a Type I hood over cooking operations producing grease-laden vapors because of this increased fire hazard. A Type I exhaust hood and duct system is constructed to be able to handle the grease-laden vapors and provide a more substantial system in a fire situation than Type II hoods. Type I systems are constructed of 16 gage steel, joints must have a continuous liquid-tight weld, and all connections must be designed not to trap grease. Prior to being placed in service, the grease duct must pass a light test to determine if any voids, or holidays, are found in the continuous welds. Portions of the ductwork that penetrate concealed areas in the building, such as attic spaces and wall cavities, must be protected by an enclosure that provides a fire-resistance rating of at least 1-hour.

In addition to the component design for fire-resistance, commercial cooking activities under a Type I hood are required to be protected by an automatic fire-extinguishing system (Figure 6-19). These fire-extinguishing systems are typically designed to activate both automatically or manually, and they will discharge the extinguishing media onto the cooking surface, in the plenum behind the filters and into the ductwork. [Ref. 904.12]

FIGURE 6-19 Type I exhaust hoods are required to be equipped with an automatic fire-extinguishing system. *(Courtesy of CaptiveAire)*

EMERGENCY AND STANDBY POWER SYSTEMS

Emergency and standby power systems are required by the IFC and IBC to provide a reliable, second source of electric power to buildings, equipment or facilities. These systems are required in buildings that represent a significant life safety challenge, present complications to emergency responders or contain certain classes of hazardous materials. The IFC provisions for these systems specify when they are required, system installation and performance requirements based on the building's use and the inspection, testing and maintenance of the power source and its transfer switch. The requirements for design and installation of emergency and standby power systems are specified in NFPA 70, *National Electrical Code*® (NEC®); NFPA 110, Standard for Emergency and Standby Power Systems; and NFPA 111, Standard on Stored Electrical Energy Emergency and Standby Power Systems. NFPA 110 addresses backup power systems that connect to an alternative energy grid or to engine-driven generator sets, and NFPA 111 addresses stored energy sources such as stationary storage batteries. [Ref. 1203.1.3]

You Should Know

The IFC and IRC require a secondary source of power for specific systems and equipment. The type of power could be
- Emergency
- Standby
- Essential.

Emergency power and standby power are defined terms in the NEC that explain the purpose and functions of these systems. The NEC contains design requirements for emergency and standby power systems; the standard does not designate the occupancies or uses where these auxiliary power systems are required. The IFC and IBC designate the occupancy or use and type of auxiliary power system that is required. In cases where the code does not specifically indicate if standby or emergency power is required, the determination of the acceptable form of auxiliary power rests with the design professional and the fire code official. Essential power is an emergency power supply that powers essential equipment in a health care facility. In addition to exit lights and fire alarms, this could include lighting for operating rooms, ventilators and other equipment that will jeopardize patients when not operating. Table 6-4 summarizes the occupancies, buildings and uses that require emergency power, standby power, or both (Figure 6-20). [Ref. 1203.2]

TABLE 6-4 Occupancies, buildings and uses requiring emergency and standby power

Where required	Type of required system	IFC section number
Ambulatory care facilities	Essential	1203.2.1
Elevators and platform lifts	Standby	1203.2.2
Emergency alarms	Emergency	
Emergency responder radio coverage	Standby	1203.2.3
Emergency voice/alarm communication systems	Emergency	1203.2.4
Exit signs	Emergency	1203.2.5
Gas detection systems	Emergency	1203.2.6
Essential systems in Group I-2	Essential	1203.2.7
Power operated locks and doors in Group I-2	Emergency	1203.2.8
Hazardous materials	Emergency or Standby, depending on the specific material	1203.2.9 #1
Highly toxic and toxic materials	Emergency	1203.2.9 #2
Organic peroxides	Standby	1203.2.9 #3
High-rise buildings	Emergency or Standby, depending on the electrical load's function	1203.2.10
Horizontal sliding doors	Standby	1203.2.11
Hydrogen fuel gas rooms	Standby	1203.2.12
Laboratory suites	Emergency or Standby, depending on the specific hazardous materials	1203.2.13
Means of egress illumination	Emergency	1203.2.14
Membrane structures—inflation systems in permanent membrane structures	Standby	1203.2.15
Semiconductor fabrication facilities	Emergency	1203.2.16
Smoke control systems	Standby	1203.2.17
Underground buildings	Emergency or Standby, depending on the electrical load's function	1203.2.18

Emergency and Standby Power Systems 85

FIGURE 6-20 This Group A indoor sports auditorium requires emergency power because its occupant load is more than 1,000.

Emergency power systems are specified when the interruption of electrical current would produce very serious life safety or health hazards. Means of egress exit sign lighting and exit illumination, electric fire pumps and elevator car illumination in high-rise buildings are electrical loads that must be connected to emergency power circuits. The wiring generally must be kept independent of other building wiring. The NEC prohibits the connection of nonemergency power loads to the emergency power circuits.

The electrical circuits for standby and emergency power need to be reliable and dependable, especially during a fire situation. The expectation is that the wiring in these circuits will survive for a minimum of 2 hours. Electrical power feeder-circuit equipment and wiring for emergency power systems require protection using a passive form of fire resistance. One of the following methods must be used to protect the electrical system as the conductors traverse the building:

1. Protected by construction with a 1-hour fire-resistance rating
2. Protective systems with a 1-hour fire-resistance rating
3. Critical circuits must comply with UL 2196, Standard for Fire Test for Circuit Integrity of Fire-Resistive Power, Instrumentation, Control, and Data Cables (Figure 6-21). [Ref. 1203.3]

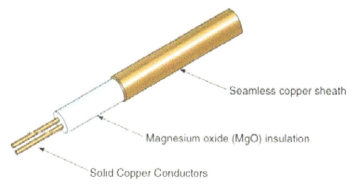

FIGURE 6-21 Pyrotenax MI cable—2-hour fire-rated cable *(Courtesy of Pentair Thermal Management, Houston, TX)*

Requirements for standby power systems are not as rigorous when compared to the emergency power requirements. The IFC requires standby power systems when the interruption of normal electrical service could create hazards or interrupt fire-fighting or rescue operations (Figure 6-22). Electrical loads required to be connected to standby power branch circuits include smoke control

86   Chapter 6 Building Systems

FIGURE 6-22 A platform lift used as part of an accessible means of egress provided with a standby power source

> **Code Essentials**
>
> Emergency power systems are required when the loss of electrical service could result in serious life safety or health hazards. Standby power systems are required when the loss of electric power could create hazards to or interrupt fire-fighting or rescue operations. ●

systems, elevators, platform lifts used as components of an accessible means of egress, and refrigeration equipment to control the storage temperatures of certain organic peroxides. Fire-resistance protection of the equipment providing standby power is not required by the NEC.

Emergency and standby power systems consist of wiring, equipment controls and an energy source that provides safe and reliable electric power (Figure 6-23). The energy source can be stored or it may be an indoor or outdoor generator set. The other critical component in an emergency or standby power system is the automatic transfer switch (ATS). The ATS is a listed electric switch that transfers the connected emergency or standby power circuits to the auxiliary energy source in the event normal electric power is disconnected. It is required to be listed for either emergency or standby power service and must be designed to safely carry the entire electrical load of all the circuits to which it is connected. An ATS must be designed so it can be manually switched between utility power and auxiliary power in case the automatic function of the switch fails when a power source

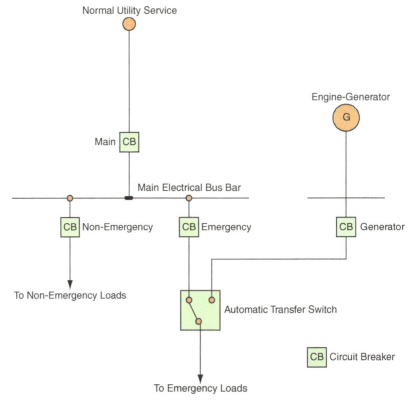

FIGURE 6-23 One-line diagram of an engine-driven generator and automatic transfer switch used as an emergency power source

transfer is required. The code requires an ATS used for emergency power service to operate within 10 seconds after power is disconnected and within 60 seconds when the switch is serving a standby power source, which is consistent with the IFC requirements for ATS devices in high-rise and underground buildings. [Ref. 1203.1.4]

Failures of emergency and standby power systems generally result from a lack of testing and exercising of the ATS or of the energy source, such as scheduled load testing of an engine-driven generator. The IFC requires that an approved maintenance and testing schedule be prepared and implemented to ensure that these systems are available for service in the event utility power is lost. Such a schedule must comply with the inspection and testing requirements in the NFPA 110 and NFPA 111. The results of inspections and tests must be documented and be maintained available for review by the fire code official. The maintenance and inspection program must include the ATS. [Ref. 1203.4]

The fire code requires that the power source, such as an engine-driven generator or a battery group, be tested under load in accordance with NFPA 110 or NFPA 111. Load testing is beneficial to engine-driven generators because during the first 12 to 24 months, full or partial loading of the engine helps to set piston rings, which is important to the overall performance of the generator under a full electrical load (Figure 6-24). If the engine piston rings are not completely set and the generator is subjected to a full design load, the engine performance can decay, causing loss of power to one or more critical loads. The tests prescribed in the IFC require tests of the primary auxiliary power source and its ATS. As an option, the standby or emergency power system can be used for load shaving. Load shaving is a practice where utility power is disconnected to a part or all of a building and a generator is used to supply electricity. Load shaving is considered equal to load testing and can be used to demonstrate compliance with the maintenance plan. [Ref. 1203.5, Exception]

FIGURE 6-24 Exercising the automatic transfer switch and load testing of an engine-driven generator are important steps in maintaining emergency and standby power systems.

EMERGENCY LIGHTING

Exit signs and emergency lighting are commonly provided with batteries as the secondary power supply (for further discussion on required exit signs and egress path illumination see Chapter 11 in this book). The batteries are typically contained within the lighting unit itself. When the main power supply is lost, the battery provides the emergency power. Battery life can shorten over the life of the equipment, so maintenance of the egress lighting is critical in providing

88 Chapter 6 Building Systems

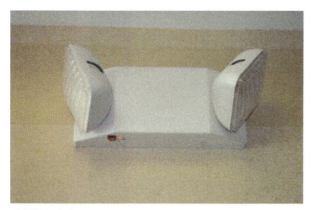

FIGURE 6-25 The button on the bottom of this emergency egress lighting unit is used to test the battery and operation of the unit.

FIGURE 6-26 Activation of the rapid shutdown switch deenergizes the system beyond 1 foot of the array boundary.

FIGURE 6-27 Solar photovoltaic disconnect switches.

for a safe evacuation of the building. The IFC requires that exit lighting is inspected and tested on a routine basis. The devices with battery backup have a test button that drops the normal power and switches power to the battery. On a monthly basis, each emergency lighting fixture must be tested for a 30-second period (Figure 6-25). Annually, these same devices must be tested for the full 90-minute duration which they are required to operate. Recordkeeping of the monthly and annual tests is required. [Ref. 1031.10]

SOLAR PHOTOVOLTAIC POWER SYSTEMS

Solar photovoltaic power systems are becoming more commonplace. The ability to convert light energy into electricity offers a cost savings to the consumer, takes advantage of a renewable resource and reduces the strain on utility power systems. These systems generate electricity that is a potential ignition source, and when installed on buildings these systems can also cause difficulty for fire-fighting operations.

These systems will generate electricity when exposed to sunlight, and the only disconnect is after the electricity is generated. For this reason, newer systems are provided with a rapid shutdown switch. This switch deenergizes most of the system outside of the solar panel array. A sign is required at the switch to identify the portions of the solar system which are shutdown (Figure 6-26). [Ref. 1204.5]

The NEC requires that the disconnect means is in a readily accessible location. NEC Section 690.13 requires the disconnect means to be marked. If the disconnect is not at the main electrical disconnect for the building, it must be visible from the point where the main service disconnect is operated (Figure 6-27). [Ref. 1204.1]

The IFC contains requirements for solar photovoltaic systems on commercial and industrial buildings, ground-mounted systems and systems on Group R-3 occupancies not constructed under the

International Residential Code (IRC). The IRC contains requirements for solar photovoltaic systems on dwellings constructed under that code (Figure 6-28).

The solar panel array area is limited in size to 22,500 square feet with a maximum dimension of 150 feet. Additionally, the solar panels must be spaced to provide fire-fighter walkways of 3 feet in width for residential buildings and 6 feet in width for commercial buildings. Spacing from roof ridges is required to provide areas for openings in the roof to be cut for fire-fighting operations (Figure 6-29). [Ref 1204.3]

FIGURE 6-28 Solar panels on a single-family dwelling are regulated under the IRC.

STATIONARY FUEL CELL POWER SYSTEMS

One method of providing secondary power is with a fuel cell. Fuel cell technology is based on generating electricity through a chemical reaction. Hydrogen is the basic fuel, but fuel cells also require oxygen. One great appeal of fuel cells is that they generate electricity with very little pollution—much of the hydrogen and oxygen used in generating electricity ultimately combine to form a harmless byproduct, namely water.

Hydrogen is either stored in a vessel for the fuel cell operation, or hydrogen can be obtained from a gas such as natural gas.

The IFC regulates fuel cells as a prepackaged system, a pre-engineered system or a field-fabricated system. In each case, the installation must meet the NEC and NFPA 853, Standard for the Installation of Stationary Fuel Cell Power Systems. [Ref. 1205.3, 1205.4]

Fuel cells are prohibited in Groups R-3 and R-4 occupancies and in dwelling units in Group R-2 occupancies. Where installed indoors, the fuel cell must be located in a room separated by 2-hour fire-resistance-rated construction in Groups A, E, I and R occupancies. In other occupancies, the separation is only required to be 1-hour fire-resistance-rated construction. [Ref. 1205.6]

A manual shutoff valve is required within 6 feet of the fuel cell, however when the fuel cell is located indoors the valve must be located in a different room unless approved by the code official. [Ref. 1205.10]

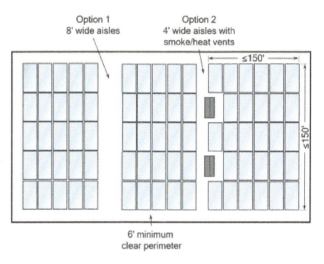

FIGURE 6-29 Solar panels on commercial buildings are limited to 150 feet by 150 feet in a single array. There are two options for aisle widths to separate the arrays.

ENERGY STORAGE SYSTEMS

Many businesses rely on computer power to run day-to-day operations, and the dependence on the electrical power supply to those computers can be critical. In an effort to maintain their computer networks in full-time operation, back-up power systems are provided. Battery back-up power supplies and capacitor storage systems are available, and as long as they are used in accordance with their listing, they meet the code requirements. However, in many businesses a separate room is needed to for the battery or capacitor system with adequate power for a multitude of computers and equipment (Figure 6-30).

FIGURE 6-30 This outdoor fuel cell is surrounded with guard posts to protect it from vehicular impact.

Additionally, the U.S. Department of Energy is working with a wide range of stakeholders to encourage the development of large scale electrical energy storage systems (ESS). ESS are needed because the amount of electricity that can be generated on the electrical grid is relatively fixed over short periods of time, although demand for electricity fluctuates throughout the day. The intent of this technology is to store electrical energy so it can be available to meet demand as needed. A number of energy storage technologies are being pursued, including battery storage systems and electrochemical capacitors, among others.

Stationary storage battery systems

With ESS technologies growing, the IFC contains regulations for known battery types and allows for future battery technologies. The types of batteries specifically regulated are
- Flow batteries
- Lead-acid batteries

- Lithium batteries
- Nickel-cadmium batteries
- Sodium batteries. [Ref. 1206.2]

The IFC contains requirements for spill control, ventilation, smoke detection systems, means to neutralize a spill, signs and markings, seismic restraint and management of thermal runaway. The requirements vary based on the specific type of battery utilized. For example, some contain liquid sulfuric acid that will require spill control and a method to neutralize a spill; whereas lithium-ion batteries do not pose the same hazard and therefore do not require the same mitigation.

It should be noted that many batteries will contain hazardous materials. Since Section 1206 is specific to stationary storage battery systems, these systems will be regulated under these requirements rather than the general hazardous materials requirements. This concept applies even though some of the provisions in Section 1206 are less restrictive than the provisions applicable to hazardous materials in general. Keep in mind that the requirements for stationary storage battery systems are designed to mitigate the specific hazards presented in these systems, whereas the general hazardous materials requirements must apply to a much broader scale of hazards.

Stationary battery systems can be located either in the same room with the actual equipment they power, or they can be located in a separate room specific for the battery storage system (Figure 6-31). The IBC will consider the battery storage room as a use that is incidental, or ancillary, to the overall use or occupancy of the building. Battery rooms will be treated as an incidental use and must be separated from the remainder of the building by fire-resistance-rated construction when the battery capacity exceeds the thresholds specified in the IFC Table 1206.2. In occupancies with a potentially high life hazard, Groups A, E, I and R, a separation of 2-hour fire-resistance-rated construction is required. In other occupancies, the separation is only required to be 1-hour fire-resistance-rated construction. A smoke detection system is required in all rooms containing a stationary battery storage system. The maximum capacity of any battery array cannot exceed 50 kWh, unless the system is a preengineered system up to a maximum of 250 kWh. [Ref. 1206.2.8]

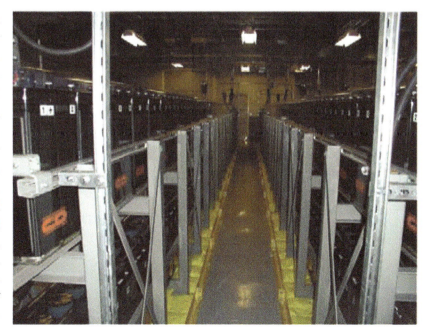

FIGURE 6-31 This battery room contains lead-acid batteries.

You Should Know

Building systems are required to ensure the safe use of a building. The provisions in IFC Chapters 6 and 12 address not only their installation but also their operation, maintenance and testing. Building systems covered in Chapter 6 can provide heated or cooled environmental air, a means for fire fighters to control elevators, a mechanism for abating hazards in a building's electricity distribution system, and a mechanical ventilation system for the removal of smoke, heat and byproducts of cooking in commercial kitchens. Building systems covered in Chapter 12 can provide electrical power utilizing solar power, emergency or standby generators, fuel cells and battery storage systems. ●

Capacitor energy storage systems

Capacitor energy storage systems (CESS) is another form of energy storage (Figure 6-32). Capacitors do not create energy, but they store energy like a battery. CESS cannot be located on a floor more than 75 feet above the lowest level of fire department vehicle access. The room containing the CESS must be separated from the remainder of the building the same as battery rooms. The maximum capacity of any capacitor array cannot exceed 50 kWh with a minimum separation of 3 feet to the next array. CESS systems located outdoors are not limited in the capacity of the array. The aggregate capacity of all CESS within a single fire area cannot exceed 600 kWh. **[Ref. 1206.3.2.3, 1206.3.3]**

FIGURE 6-32 Capacitor energy storage system *(Courtesy of Ioxus, Inc.)*

CHAPTER 7
Interior Finish and Decorative Materials

The requirements in the *International Fire Code* (IFC) and Chapter 8 of the *International Building Code* (IBC) address the selection and installation of materials, fabrics, surface treatments, furnishings and decorations installed or placed inside of buildings. Interior finishes are the surface coverings of interior walls, ceilings and floors and may include fixed or moveable wall partitions and interior paneling, wainscoting, molding or other finish that is used to decorate, insulate or reduce sound levels inside a building. Decorative materials can be natural or man-made materials that are applied over the building's interior finish for decorative or acoustical effects. Interior finish, decorative materials and furnishings become sources of fuel in a building fire. The requirements in the IFC and IBC Chapter 8 were developed after a number of structure fires resulted in hundreds of lives being lost because of exposure to smoke or because the room or area experienced a flashover.

PURPOSE OF THE REQUIREMENTS

The IBC and IFC requirements for interior finishes and furnishings were developed after a number of fires in assembly occupancies in which highly combustible materials were affixed to the interior ceiling or walls of the building, or extremely combustible decorative materials were used that contributed to rapid smoke production and fire growth. A number of major fires in the U.S. and throughout the world have resulted in hundreds of fatalities. Table 7-1 summarizes four of these major incidents. In all of the indicated buildings, the interior finish or decorative materials were a contributing factor; however, other factors including inadequate means of egress width, travel distances and improper or obstructed exit door openings also contributed to the high number of lives lost.

TABLE 7-1 Summary of large life loss fires where interior finishes or decorative materials were a contributing factor

Location	Date	Number of fatalities	Number of injuries
Rhythm Night Club, Natchez, MS	April 23, 1940	209	Unknown
Cocoanut Grove Night Club, Boston, MA	November 28, 1942	492	Unknown
Beverly Hills Supper Club, Southgate, KY	May 28, 1977	165	> 200
Station Nightclub, Warwick, RI	February 20, 2003	100	> 200

The IFC Chapter 8 requirements limit the likelihood that interior finish or decorative materials contribute to flashover. In Group I and R-2 university dormitories, the IFC also prescribes requirements that establish fire safety requirements for furnishings. The requirements limit the amount of heat released when compared to other combustible materials of equal mass and density.

Flashover has a variety of definitions. In the mind of fire fighters, a room totally involved in fire or a hot smoke layer that has visible flames are two examples of a flashover. Flashover is an event during a fire's growth where the hot smoke layer inside a room or compartment releases the greatest amount of radiant energy. During the incipient phase of fire growth, carbon monoxide (a flammable gas) and other gases and smoke particulate are produced and begin to accumulate at the top of the room, forming a hot smoke layer.

As the fire continues to progress through the pre-flashover phase, it heats the compartment including furnishings, decorative material and stored goods, which produces more heated gases

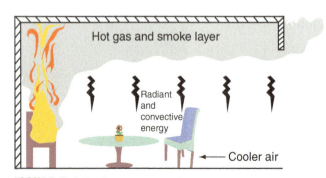

FIGURE 7-1 Preflashover conditions in a room or compartment fire

and smoke. In the incipient and pre-flashover periods of fire growth, convective heat transfer between the burning materials and the interior surfaces of the room or compartment is the dominating energy (Figure 7-1).

Additional hot gases and smoke are produced as the size of the fire increases. This in turn causes the smoke layer to thicken, increase in temperature and move closer to the combustible fuels. At this point, radiant energy becomes the dominant energy. When the temperature of the smoke layer reaches approximately 1,100°F, the level of radiant energy absorbed by the combustible materials causes them to exceed their ignition temperature and flashover occurs. Another criterion of flashover is when the radiant energy level at the floor reaches 15-20 kilowatts/square meter. These criteria at which flashover is estimated to occur will vary slightly due to different mechanisms that result from varying fuel properties, fuel orientation, geometry of the room or enclosure and conditions in the hot smoke layer.

The likelihood of surviving flashover is almost nonexistent unless the individual is wearing structural fire-fighting clothing and respiratory protection. As more combustible finishes and decorations are added to a room, the time lapse from fire ignition to flashover shortens. For this reason, it is important that the amount and type of combustible materials allowed in buildings be regulated.

The IFC and IBC provisions limit the likelihood of interior finish and decorative materials contributing to flashover by prescribing the use of flame-resistant materials with a minimum density to slow the rate of fire spread and heat release (Figure 7-2). The orientation and installation of interior finish materials and decorative materials on room walls and ceilings are also regulated, as these have a profound effect on the time to flashover. Prescribed tests are performed to ensure the material either will not cause a flashover or will have a limited ability to contribute fuel to a flashover, based on the test criterion. Flame resistance also can be provided by a thermal barrier, such as placing gypsum wallboard over expanded foam plastic.

> **Code Essentials**
>
> A compartment fire has four phases of growth and decay:
> 1. incipient
> 2. pre-flashover
> 3. flashover
> 4. post-flashover.
>
> Flashover occurs at approximately 1,100°F and is driven primarily by radiant energy within the fire's hot gas and smoke layer.

FIGURE 7-2 Synthetic foam plastic decorations suspended from a building's ceiling are in violation of IFC Chapter 8.

INTERIOR WALL AND CEILING FINISH AND TRIM

The IFC requirements for interior finish materials installed on walls and ceilings are based on their rate of flame spread and smoke production, or their ability to contribute to flashover. Different fire test methods are permitted by the IFC, and each test has its own criteria for assessing the relative fire risks of materials being evaluated. **[Ref. 803.1]**

A fire test that has been used for over 70 years is the Steiner Tunnel Test, named after Edward Steiner, an engineer at Underwriters Laboratories who developed the test apparatus. The test in now referred to in ASTM E84, Standard Test Method for Surface Burning Characteristics of Building Materials. The test involves comparing the performance of the material being evaluated to two other materials that have predictable and reproducible results. The test results are reported as a material's flame spread index (FSI) and smoke-developed index (SDI). FSI and SDI are comparative measures, expressed as a dimensionless number, derived from visual measurements of the spread of flame or smoke obscuration versus time for a material tested in accordance with ASTM E84 or UL Standard 723, Test for Surface Burning Characteristics of Building Materials (Figure 7-3). **[Ref. 803.1.2]**

FIGURE 7-3 The Skinner Tunnel Test referred to in ASTM E84 measures a material's flame spread and smoke-developed indexes. *(Courtesy of Underwriters Laboratories, Northbrook, IL)*

Flame spread is assessed visually by the progression of the flame front, while measurements of optical smoke density at the tunnel outlet determine the smoke obscuration. This information is used to plot time-based graphs of flame-spread distance and optical density. FSI and SDI are then calculated based on the ratio between the duration for the material being tested and those for asbestos cement board (assigned FSI and SDI values of 0) and for red oak flooring (assigned FSI and SDI values of 100). Based on the FSI and SDI values, the IFC assigns a class rating for interior finish materials materials as indicated in Table 7-2. The class ratings are used to assign minimum interior finish requirements inside buildings. **[Ref. 803.1.1]**

TABLE 7-2 Interior finish material classifications in accordance with ASTM E84

Material class	Flame spread index	Smoke-developed index
Class A	0–25	0–450
Class B	26–75	0–450
Class C	76–200	0–450

Although the ASTM E84 test provides a reasonable basis for comparing surface flame-spread characteristics of "traditional" building materials, such as wood, the results of this test may not predict actual fire behavior for many contemporary materials, particularly textiles and plastics. Melting and dripping of these materials during the test, and testing with a ceiling-mounted sample configuration can skew test results for contemporary materials, discrediting the ASTM E84 test results.

Because of these recognized weaknesses with the ASTM E84 test, the IFC references a more-current test method, referred to as the "room-corner test," which is prescribed by NFPA 286, Standard Methods of Fire Tests for Evaluating Contribution of Wall and Ceiling Interior Finish to Room Fire Growth. Tests conducted in accordance with this standard are recognized by the IFC as alternatives to the ASTM E84 approach to flame-spread testing, because room-corner tests do a better job of simulating actual fire conditions. In a room-corner test, materials are mounted to the walls and/or ceiling of a test room, and the fire exposure is generated by a gas burner that simulates a trash-can fire extending to a chair in the corner of the room (Figure 7-4). [Ref. 803.1.1]

FIGURE 7-4 A room-corner fire-test apparatus

The burner flame in the NFPA 286 test contacts the ceiling surface in the latter portion of the test, providing a substantial fire exposure to ceiling-mounted materials. Because the burner flame exposes both the wall and the ceiling in the NFPA 286 test, this test can be used for evaluation of both wall and ceiling finishes. To successfully pass the room-corner test, the material being tested must withstand a certain fire exposure and, depending on if the material will be installed on walls, ceilings or both, the flames must not extend beyond either the walls or boundary of the test compartment (Figure 7-5). The acceptance criteria are

1. During the 40 kilowatt (kW) exposure, flames do not spread to the ceiling.
2. The flames do not spread to the outer extremity of the sample on any wall or ceiling.
3. Flashover, as defined in NFPA 286, does not occur.
4. The peak heat release rate throughout the test shall not exceed 800 kW.
5. The total smoke released throughout the duration of the fire test does not exceed 1,000 square meters. [Ref. 803.1.1.1]

Requirements for interior finish materials are set forth in Table 803.3 of the IFC and Table 803.13 of the IBC (Table 7-3). These tables establish the requirements based on the occupancy classification of the building and where the interior finish material will be installed within

FIGURE 7-5 A room-corner fire test of wall and ceiling finish material

the means of egress system. Consider a Group A-2 occupancy: if the building is not protected by an automatic sprinkler system, Class A interior finish materials are required in all exit enclosures, exit passageways and corridors, while all rooms or enclosed spaces would require the use of interior finish materials with a Class B rating. If the building is protected throughout by an automatic sprinkler system, the interior finish materials with a higher FSI can be used.

TABLE 7-3 Interior wall and ceiling finish requirements by occupancy[k] (IFC Table 803.3)

Group	Sprinklered[l]			Nonsprinklered		
	Interior exit stairways and interior exit ramps and exit passageways[a, b]	Corridors and enclosure for exit access stairways and exit access ramps	Rooms and enclosed spaces[c]	Interior exit stairways and interior exit ramps and exit passageways[a, b]	Corridors and enclosure for exit access stairways and exit access ramps	Rooms and enclosed spaces[c]
A-1, A-2	B	B	C	A	A[d]	B[e]
A-3[f], A-4, A-5	B	B	C	A	A[d]	C
B, E, M, R-1, R-4	B	C[m]	C	A	B	C
F	C	C	C	B	C	C
H	B	B	C[h]	A	A	B
I-1	B	C	C	A	B	B
I-2	B	B	B[h,i]	A	A	B
I-3	A	A[j]	C	A	A	B
I-4	B	B	B[h,i]	A	A	B
R-2	C	C	C	B	B	C
R-3	C	C	C	C	C	C
S	C	C	C	B	B	C
U	No restrictions			No restrictions		

For SI: 1 inch = 25.4 mm, 1 square foot = 0.0929m².

a. Class C interior finish materials shall be allowed for wainscoting or paneling of not more than 1,000 square feet of applied surface area in the grade lobby where applied directly to a noncombustible base or over furring strips applied to a noncombustible base and fireblocked as required by Section 803.11.1 of the *International Building Code*.
b. In exit enclosures of buildings less than three stories in height of other than Group I-3, Class B interior finish for nonsprinklered buildings and Class C for sprinklered buildings shall be permitted.
c. Requirements for rooms and enclosed spaces shall be based upon spaces enclosed by partitions. Where a fire-resistance rating is required for structural elements, the enclosing partitions shall extend from the floor to the ceiling. Partitions that do not comply with this shall be considered as enclosing spaces and the rooms or spaces on both sides shall be considered as one. In determining the applicable requirements for rooms and enclosed spaces, the specific occupancy thereof shall be the governing factor regardless of the group classification of the building or structure.
d. Lobby areas in Group A-1, A-2, and A-3 occupancies shall not be less than Class B materials.
e. Class C interior finish materials shall be allowed in Group A occupancies with an occupant load of 300 persons or less.
f. In places of religious worship, wood used for ornamental purposes, trusses, paneling, or chancel furnishing shall be allowed.
g. Class B material is required where the building exceeds two stories.
h. Class C interior finish materials shall be allowed in administrative spaces.
i. Class C interior finish materials shall be allowed in rooms with a capacity of four persons or less.
j. Class B materials shall be allowed as wainscoting extending not more than 48 inches above the finished floor in corridors.
k. Finish materials as provided for in other sections of this code.
l. Applies when the vertical exits, exit passageways, corridors, or rooms and spaces are protected by an approved automatic sprinkler system installed in accordance with Section 903.3.1.1 or 903.3.1.2.

FOAM PLASTICS

Many interior trim and architectural components are readily available manufactured of foam plastic materials. Typical of plastics, these materials need to be regulated to limit the fuel load. However, foam plastics typically burn with a greater intensity than solid plastics because the foam plastic has a larger surface area compared to the same weight of solid plastic, and the foam plastic itself can consist of air pockets that allow air to mix easier with the fuel. The IFC has specific limitations for foam plastic materials used as trim within a building. The foam plastic pieces must have a minimum density 20 pounds per cubic foot; the heavier the density, the less air is entrained into the product. The cross-sectional dimensions cannot exceed $1/2$ inch by 8 inches. The product can have a flame spread index of up to 75. Each component must meet these criteria, and the aggregate surface area covered by foam plastic cannot exceed 10 percent of each wall. [Ref. 804.2]

> **Code Essentials**
>
> Interior wall and ceiling finish requirements were developed so that selected materials contribute little fuel to a fire. The requirements in the IFC are based on the FSI and SDI of a material. As an alternative to the ASTM E84 test, the IFC allows the use of a room-corner test, which offers a realistic assessment of a material's fire behavior.

UPHOLSTERED FURNITURE AND MATTRESSES

The IFC has requirements limiting the ignitability and heat release rate of furnishings and bedding in ambulatory care facilities (Group B), assisted living facilities (Group I-1, Condition 2), nursing homes and hospitals (Group I-2), detention and correctional facilities (Group I-3) and college and university dormitories (Group R-2). The requirements are concerned with furniture and bedding used in areas where individuals sleep or are undergoing patient care and are incapable of self-rescue.

FIGURE 7-6 The foam plastic trim used to create the columns and cornice around the window cannot exceed 10 percent of the wall area.

The requirements for determining the ignition resistance of upholstered furniture and mattresses in each occupancy class are based on tests that replicate ignition using a lit cigarette. At the end of the test, the length of the char is measured. If the char length is less than the IFC prescribed limits, which vary for upholstered furniture and occupancies, the furnishing or bedding can be introduced and used in sleeping areas. [Ref. 805.1.1, 805.1.2, 805.2.1, 805.2.2, 805.3.1, 805.3.2, 805.4.1, 805.4.2]

The heat release rate of furnishings in the indicated occupancies is also regulated by the IFC provisions. Furniture inside the indicated occupancies is limited to a maximum peak heat release rate of 80 kilowatts when tested in accordance with ASTM E1537, Standard Test Method for Fire Testing of Upholstered Furniture, or using an alternative test method known as California Technical Bulletin 133.

100 Chapter 7 Interior Finish and Decorative Materials

FIGURE 7-7 Upholstered furniture and mattresses require a tag or label indicating conformance to the IFC requirements for ignition resistance and maximum heat release rate.

Mattresses are limited to a maximum heat release rate of 100 kW. Test samples are also limited to a total energy release of 25 megajoules or less during the first 10 minutes of the test. **[Ref. 805.1.1.2, 805.2.1.2, 805.2.2.2, 805.3.1.2, 805.4.1.2]**

To assist in verifying compliance with the IFC, upholstered furniture and mattresses require a label issued by an approved agency to indicate conformance to the IFC cigarette-ignition resistance and maximum heat release rate requirements (Figure 7-7). **[Ref. 805.1.1.2, 805.1.2.2, 805.2.1.2, 805.2.2.2, 805.3.1.2, 805.3.2.2.1, 805.4.1.2, 805.4.2.2]**

The requirements in Section 805 limiting the heat release rate are not applicable in Groups I-1, I-2 and R-2 and ambulatory care facilities when the room or area is protected by an automatic sprinkler system designed and installed in accordance with Section 903.3.1.1.

The IFC also regulates combustible decorative materials that are located within certain occupancies. The materials must meet the flame spread index requirements for the occupancy and have restrictions in the area of wall or ceiling that can be covered. The amount of combustible curtains, draperies, fabric hangings and other similar combustible decorative materials is limited as shown in Table 7-4.

> **You Should Know**
>
> Chapter 8 in the IFC addresses materials that can create a significant fire load within a building. These materials can often be transient and change with tenants or the season.

TABLE 7-4 Combustible curtains, draperies and fabric hangings attached to walls and ceilings

Occupancy	Maximum surface area covered	
	Sprinklered	Nonsprinklered
Group A	75% [a]	10%
Groups B, M	10% [b]	10% [b]
Groups I-1, I-2, I-4	10%	10%
Group I-3	NA	NA
Group R-2 dormitories	50%	10%
Groups E, R-1	50% [c]	10%
All others	10%	Not Limited

NA = not allowed
a. Percentage only allowed in auditoriums of Group A.
b. The amount of ceiling area covered is not limited if the material meets the fire test criteria for NFPA 289 or 701.
c. Percentage only allowed in dormitories of Group R-2. **[Ref. 807.2, 807.3]**

PART IV: Fire/Life Safety Systems and Features

Chapter 8: Requirements for All Fire Protection Systems

Chapter 9: Automatic Sprinkler Systems

Chapter 10: Fire Alarm and Detection Systems

Chapter 11: Means of Egress

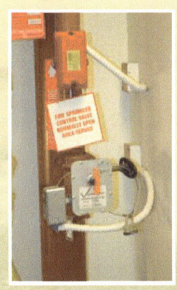

CHAPTER 8
Requirements for All Fire Protection Systems

Chapter 9 of the *International Fire Code* (IFC) sets forth requirements for active fire protection systems. A fire protection system is defined as "approved devices, equipment and systems or combinations of systems used to detect a fire, activate an alarm, extinguish or control a fire, control or manage smoke and products of a fire or any combination thereof." Given this definition, a fire protection system can include an automatic sprinkler system, an alternative automatic fire-extinguishing system, a fire pump or smoke alarms in one- and two-family dwellings. A fire protection system prescribed by the IFC must perform one or more of the following functions:
- Detect a fire
- Activate an alarm
- Extinguish or control a fire
- Control or manage smoke or other combustion byproducts of a fire. [Ref. 902.1]

Chapter 8 introduces readers to the general provisions in IFC Section 901. These provisions are applicable to all fire protection systems required by the IFC and optional systems that may be specified by the design professional. Installation of fire protection systems that are not required is allowed, provided they are installed in accordance with all of the requirements in the IFC and *International Building Code*. [Ref. 901.4.2]

WHERE ARE FIRE PROTECTION SYSTEMS REQUIRED?

The provisions requiring a fire protection system vary in the IFC and IBC. Outside of buildings, the IFC requires the installation of private fire protection water mains and fire hydrants to ensure that the required fire-flow is available in the event of a structure fire. IBC and IFC fire protection system requirements inside of buildings are based on

1. The occupancy of the building [Ref. 903.2]
2. The size of fire areas within the building [Ref. 903.2]
3. The occupant load of the building [Ref. 903.2]
4. The height or area of a building [Ref. IBC Table 504.3, IBC Table 504.4, IBC Table 506.2]
5. The quantity of hazardous materials stored or used inside a building [Ref. 5003.1.1]
6. The fire loss history of a given hazard or operation.

Certain occupancy classifications require the installation of one or more fire protection systems based on these six characteristics. In the case of a residential occupancy (Group R), an automatic sprinkler system is required throughout the building when a Group R fire area is created. For example, if part of a small one-story business (Group B) were converted to a motel, this change of occupancy requires the installation of an automatic sprinkler system throughout the entire building. Conversely, the construction of a moderate-hazard factory occupancy (Group F-1) would not require automatic sprinkler protection unless its fire area exceeds 12,000 square feet. Many of the IFC fire protection system requirements are based on the building occupancy, occupant load or fire areas. [Ref. 903.2.4, 903.2.8]

The occupant load of a building can dictate the installation of certain fire protection systems. In an assembly occupancy, such as a dance hall where food or beverages are not served (Group A-3), an automatic sprinkler system is required if the occupant load is 300 or more persons. Because of the large occupant load, the IFC prescribes the installation of occupant notification devices as part of a fire alarm system that would inform occupants in the event that the automatic sprinkler system operates (Figure 8-1). [Ref. 903.2.1.3, 907.2.1]

FIGURE 8-1 A Group A-3 occupancy such as a live music venue with an occupant load of 300 or more would require automatic sprinkler protection and a fire alarm system with occupant notification throughout the fire area.

FIGURE 8-2 The occupancy and building's construction type establish the allowable height and area. Allowable height and area can be increased by certain designs of automatic sprinkler protection.

FIGURE 8-3 The quantity of stored materials or their hazards, such as high-piled combustible storage as shown in this photograph, can mandate the installation of a fire protection system.

A building's height or area can dictate the installation of a fire protection system. The IBC limits building height and area based on its occupancy classification and construction type (Figure 8-2). IBC construction types are fire-resistive, noncombustible, combustible or a mixture of combustible and noncombustible materials. In many occupancies, the IBC permits the building's height and area to be increased when the building is protected by an automatic sprinkler system. The number and type of increases granted depends on whether the automatic sprinkler system is designed for property and life safety protection or if it is only designed for life safety applications. [Ref. 903.3.1.1]

A fire protection system may be required inside a building when the quantity of materials stored or their hazards exceed certain limits. IFC Chapter 32 regulates high-piled combustible storage, which is a very common method of storing large volumes of materials in Group S and Group M occupancies (Figure 8-3). Automatic sprinkler protection in these buildings is required based on the fire hazard of the stored goods, the storage height and area and the building's occupancy classification. Certain classes of hazardous materials are so easily ignited or can cause such a large amount of damage if ignited that IFC Chapter 50 requires the installation of an automatic sprinkler system throughout buildings when these classes of hazardous materials are stored indoors. [Ref. 903.2.7.1, Table 5003.1.1(1) Footnote g]

Certain hazards inside of buildings represent a relatively high threat to the occupants, the building and its contents if an unwanted fire occurs (Figure 8-4). To control the fire risks, the IFC requires the installation of fire protection systems for specific hazards or uses. One example is commercial cooking systems that can generate smoke and grease-laden vapors. Because the smoke and vapors act as fuel inside of the commercial cooking exhaust system, the IFC requires the installation of a fire-extinguishing system specifically designed for the hazard. The system is designed in accordance with the applicable NFPA standards and IFC installation requirements. Any modification to an existing commercial cooking system requires the fire protection system be upgraded to comply with the IFC. [Ref. 904.12, 904.12.5.1]

FIGURE 8-4 The IFC requires certain fire hazards that represent a relatively high threat to building occupants, contents and the structure itself be protected by a fire protection system. *(Courtesy of Ansul/Tyco Inc., Marinette, WI)*

A jurisdiction may be asked to approve the construction of a building, structure or process where the hazards are unique or challenging to emergency responders. In some cases, the size or arrangement of the hazard may impair or limit the ability of fire apparatus to approach the location. In such instances, the fire code official can require additional fire protection safeguards. These can be in the form of any fire protection system as addressed in Chapter 9 of the IFC. Consistency in code enforcement is important. Once the fire code official decides additional protection is required, it must also be required for any similar installations in the future; if not, the jurisdiction could be liable for the added expenses imposed on one business while waiving the requirement for others. Additional fire protection systems or safeguards may not be so obvious or rely on fire suppression and detection systems (Figure 8-5). Realize that certain process safety controls specified by the design professional, such as installing certain flow control valves, automatic shut-down components or other less obvious but equally reliable systems, may effectively reduce and manage a hazard while providing greater reliability than a fire suppression or detection system. In instances when the fire code official is considering additional fire protection systems, it's reasonable and prudent to request that the design be reviewed by a fire protection engineer or other competent design professional who has the experience and understanding of the hazards of the building, occupancy, the stored materials or the process (see Chapter 1 in this book).

[Ref. 901.4.4]

> **Code Essentials**
>
> The IFC requires fire protection systems inside and outside of buildings. When a fire protection system is required indoors, it can be based on the building's occupancy classification, occupant load, its height or area or the hazards of the stored goods and materials.

FIGURE 8-5 This additional fire protection system can be required by the fire code official. The manual master stream nozzles are used for fire exposure protection of large above-ground storage tanks storing aviation jet fuel.

CONSTRUCTION DOCUMENTS AND ACCEPTANCE TESTING

IFC Section 105.7 requires a construction permit for the installation or modification of a fire protection system. Fire protection system design includes compliance with the adopted NFPA standards for the specific systems in addition to any design criteria in the IFC. Table 8-1 lists the fire protection systems and standards adopted by the IFC. [Ref. 901.6.1]

TABLE 8-1 Fire protection systems and standards

Fire protection system or component	NFPA standard
Low-, Medium-, and High-expansion Foam	11
Carbon Dioxide Extinguishing Systems	12
Halon 1301 Fire Extinguishing Systems	12A
Sprinkler Systems	13
Sprinkler Systems in One- and Two-Family Dwellings and Manufactured Homes	13D
Sprinkler Systems in Low-Rise Residential Occupancies	13R
Standpipe and Hose Systems	14
Water Spray Fixed Systems for Fire Protectection	15
Foam-Water Sprinkler and Foam-Water Spray Systems	16
Dry-Chemical Extinguishing Systems	17
Wet-Chemical Extinguishing Systems	17A
Stationary Pumps for Fire Protection	20
Water Tanks for Private Fire Protection	22
Private Fire Service Mains and Their Appurtenances	24
Fire Alarm Systems	72
Smoke Control Systems	92
Ventilation Control and Fire Protection of Commercial Cooking Operations	96
Smoke and Heat Vents	204
Water Mist Fire Protection Systems	750
Water Supply for Suburban and Rural Fire Fighting	1142
Clean Agent Fire Extinguishing Systems	2001
Fixed Aerosol Fire-Extinguishing Systems	2010

The IFC commonly requires more than one fire protection system in certain buildings. A building with an occupied floor more than 75 feet above the lowest level of fire department access is defined as a high-rise building. The IBC requires not only an automatic sprinkler system, but also a standpipe system, a fire alarm and detection system equipped with an emergency voice/alarm communication system, a generator to provide standby and emergency power and possibly a fire pump or smoke control system. All of these systems must comply with the IFC and NFPA requirements. [Ref. 901.4.1]

The IFC requires the permit applicant to submit construction documents to the fire code official for review and approval of the fire protection system to verify the installation or modification is properly designed for the hazards it is protecting (Figure 8-6). The scope and detail of the submitted documents are commonly specified in the NFPA standard applicable to the system design or modification. In many cases, the submittal will include manufacturer's equipment data or cut sheets and various formats of engineering calculations used as the basis for the design. [Ref. 104.2, 105.2, 105.4, 901.2, 907.1.1, 909.2]

FIGURE 8-6 Fire protection system construction documents must be submitted to the fire code official to verify that the design will protect the identified hazards and will meet the requirements of the applicable technical standards.

Prior to requesting a final approval of a fire protection system installation or modification, the fire code official can require the installing contractor to submit a statement of compliance. The statement of compliance documents that the fire protection system was installed in accordance with the applicable design standard, the construction specifications prepared by the design professional and the manufacturer's instructions for particular components. Any deviations from the installing standard are also attached to the statement of compliance. This statement may also be used to document that the fire protection system has been tested by the installing contractor—this is an important consideration, especially on larger or complicated systems. If the fire protection system is not tested and exercised prior to a final inspection, it generally fails to pass the inspection. [Ref. 901.2.1]

The NFPA standards that govern the design of automatic sprinkler systems, wet-chemical alternative fire-extinguishing systems, private underground fire protection water mains, and fire alarm and detection systems each contain an example statement of compliance. In instances where an example statement of compliance is not available, the jurisdiction can develop its own, require the installing contractor to provide his or her own statement, waive the requirement when allowed by the fire code official or use example forms available in the *Fire Plan Review and Inspection Guidelines* published by ICC (Figure 8-7). In states that license contractors, a statement of compliance may be required to be prepared by the installing contractor.

> **Code Essentials**
>
> Installation or modification of fire protection systems requires
> - The submittal of construction documents, which can include equipment cut sheets and calculations
> - A construction permit
> - A statement of compliance when it is required by the fire code official

Statement of Compliance
Stationary Fire Pump

Permit #:_____   Date:_____

	Property Protected	Installing Contractor	General Contractor
Business Name:	_____	_____	_____
Address:	_____	_____	_____
Representative:	_____	_____	_____
Telephone:	_____	_____	_____

Location of Plans: _____

Location of Owner's Manual: _____

1. Certification of System Installation: This system installation was inspected and found to comply with the installation requirements of:

_____ NFPA 20 and the National Electrical Code®
_____ Manufacturer's Instructions
_____ Other (specify; FM, UL, etc.) _____

Print Name: _____

Signed: _____ Date: _____

Organization: _____

2. Certification of System Operation: All operational features and functions of this system were tested and found to be operating properly in accordance with the requirements of:

_____ NFPA 20 and National Electrical Code®
_____ Design Specifications and Submitted Shop Drawings
_____ Manufacturer's Instructions
_____ Other (specify) _____

Print Name: _____

Signed: _____ Date: _____

Organization: _____

FIGURE 8-7 Stationary fire pump statement of compliance

INSPECTION, TESTING AND MAINTENANCE

Fire protection systems require mechanical and electrical components to properly operate and to perform their intended functions. Water-based fire protection systems must be connected to a reliable source of water and alternative automatic fire-extinguishing systems require a sufficient volume of wet, dry or gaseous agent to extinguish a fire. Fire protection systems that discharge water or another alternative agent are constructed with listed nozzles that must not be obstructed, blocked, painted or improperly oriented. Fire protection systems must be maintained so they are available when ignition occurs. Defective components that have failed or that are inoperable must be repaired or replaced. [Ref. 901.6]

Required and nonrequired fire protection systems must be inspected, tested and maintained in accordance with the applicable fire protection system standards (Table 8-2). For water-based fire protection systems including private water storage tanks, fire hydrants supplied from private fire protection water mains, automatic sprinkler systems, standpipes and fire pumps, the IFC adopts NFPA 25, Standard for the Inspection, Testing, and Maintenance of Water-Based Fire Protection Systems. Most water-based fire protection systems require an annual inspection and one or more tests. The IFC prescribes increased inspection intervals for alternative automatic fire-extinguishing systems. NFPA standards for inspection, testing and maintenance are set forth in Table 901.6.1 (Table 8-3). [Ref. 901.6.1]

TABLE 8-2 Fire protection system inspection, testing and maintenance

Inspection	A periodic visual check to ensure the system or equipment is in place, is not impaired and appears ready for operation.
Testing	A functional or performance test by a qualified person to verify that the system or equipment will operate as intended.
Maintenance	Periodic service performed by a qualified person in accordance with the system or equipment manufacturer's recommendations to extend operational life and to enhance reliability.

TABLE 8-3 Fire protection system maintenance standards (IFC Table 901.6.1)

System	Standard
Portable fire extinguishers	NFPA 10
Carbon dioxide fire-extinguishing systems	NFPA 12
Halon 1301 fire-extinguishing systems	NFPA 12A
Dry-chemical extinguishing systems	NFPA 17
Wet-chemical extinguishing systems	NFPA 17A
Water-based fire protection systems	NFPA 25
Fire alarm systems	NFPA 72
Smoke and heat vents	NFPA 204
Water-mist systems	NFPA 750
Clean-agent extinguishing systems	NFPA 2001
Aerosol fire-extinguishing systems	NFPA 2010

Records of inspection, testing and maintenance must be maintained on the premises for at least three years (Figure 8-8). Upon request, these records must be copied to the fire code official. In addition, the initial installation records for the fire protection system must be maintained at the site. The records must include the name of the installing contractor, equipment cut and data sheets and the manufacturer's installation and instruction manuals. [Ref. 901.6.3]

FIGURE 8-8 Fire protection systems must be inspected, tested and maintained.

When buildings contain multiple fire protection systems that interact, there is a need for integrated testing of these systems. A simple example is a fire sprinkler operates, the water flow switch is activated, a local alarm sounds and a signal is sent to a monitoring location. This typical installation consists of multiple systems operating in sequence, and if any item fails, the sequence is broken. More complex systems are being installed in buildings involving multiple systems and, as a result, multiple contractors. For example, a smoke detector initiates a signal to capture elevators, close doors, open or close dampers, start the smoke control system in the atrium and activate fans to pressurize the elevator shafts. These systems must be tested at time of acceptance and throughout the life of the building to ensure that all of these various systems and control panels operate properly. The code references NFPA 4, Standard for Integrated Fire Protection and Life Safety System Testing for handling integrated testing of the complex systems. When integrated testing is required, it must be accomplished at the time of acceptance and at least every 10 years thereafter. [Ref. 901.6.2]

FIRE PROTECTION SYSTEM IMPAIRMENT

When a fire protection system is impaired, the fire department and fire code official must be notified. When a system is out of service, the fire code official has the authority to require evacuation of the building or require that the continued use of the building is under the supervision of a fire watch. After the fire protection system is returned to service, the fire watch is discontinued. [Ref. 901.7]

FIGURE 8-9 This wet-pipe automatic sprinkler system has a closed water supply valve. Note the lack of pressure on the gauges.

Impairments of fire protection systems can be categorized into two broad categories: unscheduled or scheduled. An unscheduled impairment includes the loss of utility electricity to an electric fire pump or the failure of a water main supplying an automatic sprinkler system, or it could be an individual turning off a water supply valve to a sprinkler system because a fire sprinkler was hit by a forklift. A scheduled impairment includes testing, maintenance or modifications to a fire protection system (Figure 8-9). For example, if a tenant improvement is constructed in a covered mall building where several lease spaces are consolidated into a single retail space, automatic sprinkler system modification is an anticipated and foreseeable event. When a fire protection system is impaired, either the building owner or a designated employee assumes the role of impairment coordinator. An impairment coordinator is responsible for the maintenance of the fire protection system. [Ref. 901.7.1]

The impairment coordinator must perform several actions before authorization can be granted to remove the system from service, including the following:

1. Identify the extent and duration of the impairment.
2. Inspect the areas of the building that will be affected by the impairment and identify any processes or hazards that need to be discontinued. In some cases, this may require prescribing the use of less hazardous operations such as using hand tools rather than machinery.
3. Notify the fire department, the approved fire alarm system monitoring station, as well as the required notifications within the organization's hierarchy, such as the loss control department, the safety coordinator and the insurance underwriter.
4. Assemble the necessary tools and materials to perform the modification or repair.
5. Implement the impairment tag program. [Ref. 901.7.4]

After the impairment is concluded, the impairment coordinator must notify the organizations and individuals specified in Section 901.7.4 and ensure that that the fire protection system is inspected and tested to verify it is operational and available for service. [Ref. 901.7.6]

When impairment occurs, a tag is required at specified locations to indicate the fire protection system is out of service (Figure 8-10). These locations include the fire department connection, system control valves, fire alarm control unit, fire alarm annunciator and, if one exists, the fire command center. Once the system has been returned to service, the impairment tag is removed. [Ref. 901.7.2, 901.7.3]

All fire protection systems are subject to these maintenance requirements including automatic sprinkler systems, fire alarm systems, fire-extinguishing systems, smoke control systems and the fire command center. Regardless of the type of fire protection system, it must be designed to protect the occupants and the building. When any of these systems are nonoperational, it is critical that notifications occur, the system is repaired and the system returned back to operational service. The impairment tag is a reminder that the system is out of service and actions must be taken to correct the situation (Figure 8-11). [Ref. 901.7, 901.7.2]

FIGURE 8-10 This impairment tag attached to the closed sprinkler control valve indicates that the system is down for service or repair.

Chapter 8 Requirements for All Fire Protection Systems

```
No.: 11500
```

Fire Protection System or Component Impairment

Removal of this tag shall be authorized by the Impairment Coordinator

Address: _____

Building No.: _____   Floor Level: _____

Fire Protection System Impaired

☐ Automatic Wet Pipe Sprinkler System - System Number _____
☐ Automatic Dry Pipe Sprinkler System - System Number _____
☐ Automatic Pre-Action Sprinkler System - System Number _____
☐ Alternative Automatic Fire Extinguishing System - System Number _____
☐ Automatic Fire Alarm System - System Number _____
☐ Kitchen Hood Fire Suppression System - System Number _____
☐ 1,500 GPM Electric Fire Pump   ☐ Fire Door - Number _____

Impairment Information

Description of Problem: _____

Reported By: _____   DATE: [ ][ ][ ][ ][ ][ ]

Restoration to Service

Summary of Repairs: _____

Repaired By: _____   DATE: [ ][ ][ ][ ][ ][ ]

☐ 2-inch Main Drain Flow Test Performed Static P _____ Residual P _____
☐ Alarm Initiating Device Test and Alarm Signal Verification

Impairment Coordinator Verification

Observed By: _____   DATE: [ ][ ][ ][ ][ ][ ]

☐ Restoration Approved   ☐ Restoration Not Approved

FIGURE 8-11 Example of an impairment tag

FIRE PROTECTION SYSTEM MONITORING

In most occupancies, the IFC requires monitoring of automatic sprinkler systems and fire alarm systems. Monitoring includes detecting and reporting alarm conditions and potential impairments, such as closed water supply valves or electrical faults, and is required to ensure a timely response by the fire department. Monitoring is not required for one- and two-family dwellings and townhouses with automatic sprinkler systems required by the *International Residential Code* (IRC) provisions. [Ref. 903.4, 907.6.6]

Automatic sprinkler system monitoring is electric supervision of valves that control the water supply to the system and water flow alarms or pressure switches that operate when a sprinkler is activated (Figure 8-12). Upon activation, these devices transmit a signal to a fire alarm control unit. The fire alarm control unit in turn transmits a signal by way of telephony, the Internet or through a wireless signal to a central or proprietary supervising station. A central supervising station is a third-party service that receives signals from fire and security systems, processes the signals and notifies the fire department and building owner of an alarm activation. A proprietary supervisory station is one owned by the property being protected, such as a campus police department. [Ref. 903.4.1]

With the exception of smoke alarms in one- and two-family dwellings and correctional and detention facilities (Group I-3 occupancies), all required fire alarm and detection systems required by the IFC must be electrically supervised. [Ref. 907.6.6]

At the time of initial installation and approval, the monitoring service should be inspected and tested. Throughout the life of a building, the monitoring service could be provided by several different companies. The building owner can switch to a different company any time he or she desires, or the owner can simply cancel the contract with the monitoring agency. The IFC requires that monitoring be provided and maintained throughout the life of the building. When a business owner cancels the service provided by the supervising station, the monitoring service provider is required to notify the fire code official in writing. This provides notice to the fire code official and allows the inspector to follow up with the facility to ensure that monitoring service is continued through another provider. [Ref. 901.9]

You Should Know

The intent of the IFC fire protection system requirements:

- Systems are designed and constructed in accordance with the applicable NFPA standards.
- Systems are designed to protect the hazards inside the building.
- Signals from most required automatic sprinkler and fire alarm systems are transmitted to a central monitoring station, which in turn notifies the fire department.
- Systems are inspected, tested and maintained in accordance with the IFC and the applicable NFPA standards.

FIGURE 8-12 This indicating floor control valve and water-flow switch is electrically supervised and monitored.

CHAPTER 9
Automatic Sprinkler Systems

Automatic sprinkler systems are the most reliable fire protection system. They are designed to detect, report and control a fire until the fire department arrives to suppress it. An automatic sprinkler system provides a means of fire detection, because sprinklers are constructed with a heat-sensitive element that operates within a specific temperature range during the early stage of fire growth. When heated above a preset temperature, the sprinkler fusible link or frangible bulb operates releasing a dispersed spray of water directly onto the fire. The discharged water absorbs the heat produced by a fire and cools the air in the room or around the fire reducing the amount of heat released and wets the unburned fuel ahead of the fire slowing its rate of fire spread. The water flow in the sprinkler system will be detected at the sprinkler riser and an electric signal will be transmitted to a supervising station which in turn notifies the fire department.

The *International Building Code* (IBC) and the *International Fire Code* (IFC) offer many credits for buildings protected by an automatic sprinkler system. Under the IBC height and area provisions for certain occupancies—a building's height can be increased by one story and the allowable area increased up to 300 percent when it is protected by an automatic sprinkler system. The IFC reduces fire-flow requirements to 25 percent when buildings are sprinklered. These are just three of the many fire protection modifications allowed by the I-Codes. Because many of these modifications can influence the ability of the fire department to control a fire if the automatic sprinkler system does not perform its function, it is important that the design and installation of the automatic sprinkler system comply with the applicable IFC and NFPA requirements.

LEVEL OF EXIT DISCHARGE AND FIRE AREA

When applying the IFC and IBC automatic sprinkler system requirements, two prerequisite terms must be understood: level of exit discharge and fire area. Proper application of these two terms, along with the building occupancy, is essential in determining where automatic sprinkler systems are required.

The level of exit discharge is defined as the story at the point at which an exit terminates and an exit discharge begins. The level of exit discharge is the floor level that allows occupants to leave the building as part of the egress path. A building can have more than one level of exit discharge (Figure 9-1). All travel that is outside the building is considered as part of the exit discharge until the person reaches the public way. At this point, exiting is complete and the person is deemed safe under the means of egress provisions in the IFC. [Ref. 1002.1]

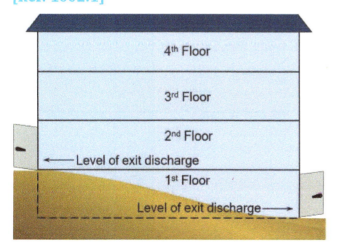

FIGURE 9-1 This building has two levels of exit discharge. The 2nd floor is the highest level of exit discharge, while the 1st floor is the lowest level of exit discharge.

> **Code Essentials**
>
> The IFC and IBC requirements for an automatic sprinkler system are based on
> - The occupancy classification
> - Fire area or building area
> - Location in relation to level of exit discharge
> - The occupant load of the fire area
> - The use conducted in the fire area
> - The amount of combustible or hazardous materials

A building's fire area is "the aggregate floor area enclosed and bounded by fire walls, fire barriers, exterior walls or horizontal assemblies of a building." Areas of the building not provided with surrounding walls shall be included in the fire area if such areas are included within the horizontal projection of the roof or floor next above (Figures 9-2, 9-3 and 9-4). A fire area is a means of creating a compartment of a given area to limit the spread of fire in a building. By using fire-resistive construction, compartments are created to contain the spread of fire for a period of time. By definition, a "fire area" is the aggregate floor area enclosed and bound by fire walls, fire barriers, exterior walls or fire-resistance-rated horizontal assemblies of a building. By separating fire areas using fire-resistance-rated construction, separate fire compartments are created in the building. When evaluating buildings regarding the need for fire sprinkler systems, each fire area of a building can be considered at separate risk during a fire incident. A building of combustible or noncombustible construction can be divided into multiple fire areas. **[Ref. 202]**

The walls or floor/ceiling assemblies separating fire areas must be fire-resistance-rated construction, and the fire-resistance rating must be at least 1 hour. But many times, the required rating will exceed 1 hour. If the building is designed as a separated mixed occupancy, the requirements in IBC Table 508.4 for separating mixed occupancies range from 1-hour to 4-hour separations. Often times, the building owner chooses to create fire areas within a single occupancy to eliminate the need to install an automatic sprinkler system. When this design is chosen, the IFC refers to IBC Table 707.3.10 for the required fire-resistance rating of the separations (Figure 9-5). **[Ref. 901.4.3]**

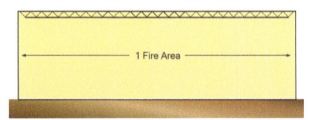

FIGURE 9-2 Each building is at least one fire area surrounded by exterior walls, floor and roof.

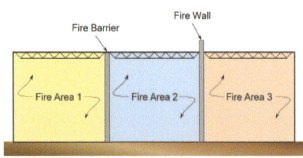

FIGURE 9-3 This 1-story building has three fire areas—two separated by a fire barrier and two separated by a fire wall.

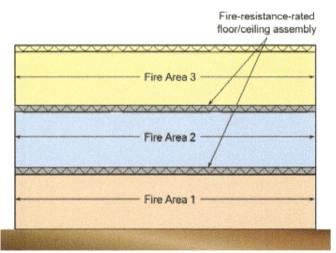

FIGURE 9-4 This 3-story building has three fire areas separated by fire-resistance-rated floor/ceiling assemblies.

Given: 18,000-square-foot building with one tenant:
  11,000-square-foot fire area
  7,000-square-foot fire area
  Ocupancy classification is Group S-1
  No high-piled combustible storage

Determine: Separation requirements for creating two fire areas to eliminate automatic sprinkler sytem requirement.

Solution: With a 3-hour fire barrier, each fire area ≤ 12,000 square feet

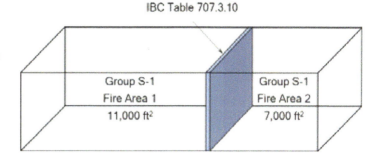

FIGURE 9-5 Application of the fire area concept

The fire area methodology set forth in the IBC, applicable only in limited occupancy groups under limited conditions, allows for the omission of automatic sprinkler protection[3]. Using fire-resistive construction to create multiple fire areas in a single building is a design alternative to the requirements for an automatic sprinkler system. Other fire protection system requirements, such as fire alarm and detection systems, are not prescribed using the fire area method. When a fire wall or fire barrier is used to compartmentalize or separate a structure, only the fire area that exceeds the area limits for the occupancy requires automatic sprinkler protection unless otherwise specified in the code. There are occupancies where the IFC will specify that all fire areas are aggregated, or that the fire-resistive separations do not eliminate fire sprinklers. The requirements for each occupancy must be reviewed in each case.

DESIGN AND INSTALLATION STANDARDS

The standards governing the design of the automatic sprinkler system required by the IFC for the protection of occupancies or buildings will influence the design of the building. The IFC adopts three NFPA standards that address the design and installation of automatic sprinkler systems:

- NFPA 13, Standard for the Installation of Sprinkler Systems
- NFPA 13R, Standard for the Installation of Sprinkler Systems in Low-Rise Residential Occupancies
- NFPA 13D, Standard for the Installation of Sprinkler Systems in One- and Two-Family Dwellings and Manufactured Homes

Automatic sprinkler systems for life safety and property protection are designed in accordance with the requirements of NFPA 13.

---

3. Thornburg, Douglas and John Henry, *2012 International Building Code Handbook,* International Code Council, Washington D.C., pg. 285.

Depending on the occupancy, automatic sprinkler systems whose primary function is for life safety are designed to either NFPA 13R or NFPA 13D. NFPA 13R systems are designed to protect Group R-1 and R-2 occupancies up to four stories in height, while NFPA 13D is used to design these systems in Groups R-3 and R-4 and townhouses. In one- and two-family dwellings and townhouses, NFPA 13D can be used as an alternative to the requirements in Section P2904 of the *International Residential Code* for dwelling fire sprinkler systems. **[Ref. 903.3.1.1, 903.3.1.2, 903.3.1.3]**

The three NFPA standards have their own specific design requirements for the type of sprinklers, water supplies and extent of protection. Table 9-1 summarizes these major considerations for each of the three standards and IRC Section P2904.

TABLE 9-1 Automatic sprinkler system design considerations

Design consideration	Sprinkler standard		
	NFPA 13	NFPA 13R	IRC Section P2904 or NFPA 13D
Extent of protection	Throughout the building (Section 903.3.1.1)	Occupied spaces (Section 903.3.1.2)	Occupied spaces (Section 903.3.1.3)
Design intent	Life safety and property protection	Life safety	Life safety
Applicability	All occupancies	Group R occupancies up to 4 stories	One- and two-family dwellings and townhouses
Design methods	Pipe schedule Control mode – density/design area Control mode – specific application Suppression modes	4 sprinklers in the hydraulic remote compartment	IRC Section P2904 - Prescriptive design; NFPA 13D - 2 sprinklers in the hydraulic remote compartment or design option identical to IRC Section P2904
Sprinklers	All listed and approved sprinklers	Listed residential sprinklers	Listed residential sprinklers
Minimum water supply duration	30 to 120 minutes depending on the hazard and design	30 minutes	IRC Section P2904 - Between 7 and 10 minutes depending on dwelling area and the number of stories; NFPA 13D - 10 minutes

With the exception of the prescriptive design methods in IRC Section P2904 and NFPA 13D Section 10.4.9 and the pipe schedule design method in NFPA 13, all other design methods are prepared using hydraulic calculations. Pipe schedule systems, which do not require hydraulic calculation, are only allowed for light- and ordinary-hazard sprinkler designs. Extra hazard sprinkler designs cannot be prepared using the pipe schedule method. Hydraulic calculations that prove the water supply is capable of providing the required volume and pressure to the sprinkler system must be a part of the plan review package submitted to the fire code official. Requirements for the preparation of these calculations are established in NFPA 13. Calculations are based on the available pressure and flow of the water supply, changes in elevation between the water supply and the sprinklers, the selected sprinklers, the NFPA 13 hazard classification for the building and the loss of pressure that results from the friction of water moving inside the pipe. Friction loss is influenced by the flow rate, the type of piping material, the pipe diameter and length and the number and type of connected fittings.

Automatic sprinkler systems are designed to either control or suppress a fire. Many sprinkler systems are designed using the control mode density/area method, which is a design intended to control the fire. Fire control is accomplished when a sufficient volume of water is discharged to control the fire and begin prewetting of exposure commodities. Prewetting slows the spread of fire. The control-mode method relies on the application of a sufficient amount of water in the right location to lower fire gas temperatures so the rate of fire growth is slowed (Figure 9-6).

Control mode specific application automatic sprinklers have a larger discharge orifice when compared to sprinklers designed for density/area applications. These sprinklers produce large droplets to provide fire control for specific high-challenge fires. These sprinklers are designed to:

1. Discharge water directly onto the fire,
2. Prewet combustibles surrounding the actual fire area to limit fire size,
3. Cool the roof area directly over the fire to prevent structural failure, and
4. Cool the roof area remote from the actual fire area so that too many sprinklers do not open and overtax the water supply.

These sprinklers differ from control mode density/area sprinklers in that a fire protection system utilizing control mode specific application sprinklers is designed based on a given number of sprinklers operating at a specified minimum water supply pressure. Control mode specific application sprinklers offer the advantage of higher discharge densities at lower water supply pressures, which can achieve adequate sprinkler protection without requiring a fire pump.

Control-mode sprinklers are designed to halt the spread of the fire; suppression-mode sprinklers are designed to provide complete extinguishment of the fire without additional hose streams from fire fighters. The design of these suppression-mode sprinklers requires that they extinguish (suppress) the fire. Suppression-mode designs commonly involve the use of early suppression fast response (ESFR) sprinklers. ESFR sprinklers bring a higher level of protection to buildings; however, they are intended for very specific uses and have very restrictive rules for installation. The installation rules must be respected over the life of the building. The performance of ESFR sprinklers can be seriously impacted by obstructions such as light fix-

Code Essentials

Residential sprinkler systems are designed to limit a fire and prevent flashover. Sprinkler systems in commercial buildings are designed to either control or suppress a fire, depending on their design, and will also prevent flashover.

FIGURE 9-6 Control-mode sprinkler *(Courtesy of TYCO Fire Suppression and Building Products, Lansdale, PA)*

Code Essentials

Sprinkler systems designed to control a fire will stop the fire spread and limit the fire damage. Sprinkler systems designed to suppress a fire will not only control the fire spread, but will also completely extinguish the fire.

FIGURE 9-7 Suppression-mode sprinkler *(Courtesy of TYCO Fire Suppression and Building Products, Lansdale, PA)*

tures or banners hanging from the structure. These sprinklers require constant evaluation, especially in speculation warehouses. Fire code officials should be cautious of building owners or design professionals who believe that ESFR sprinklers will protect all buildings from all hazards.

The expression "suppression" relates to sprinkler system performance whereby the first few sprinklers to operate provide sufficient water to the fire to reduce it promptly to an acceptable level, if not extinguish it. The effectiveness of suppression-mode sprinklers depends on the combination of fast response and the quality and efficiency of the sprinkler discharge (Figure 9-7).

TYPES OF AUTOMATIC SPRINKLER SYSTEMS

Automatic sprinkler systems are designed to discharge a given volume of water over the area being protected. Their features can vary because of the hazards of materials being stored, the lack of heat to maintain warm air inside a building so the pipe does not freeze or the potential risk of property damage resulting from accidental operation of the sprinklers.

NFPA 13 establishes design and installation requirements for four different types of automatic sprinkler systems:

- Wet-pipe automatic sprinkler system
- Dry-pipe automatic sprinkler system
- Pre-action automatic sprinkler system
- Deluge automatic sprinkler system.

Wet-pipe automatic sprinkler systems have the most reliable design because they require the least number of components (Figure 9-8). The piping supplying the sprinklers is charged with water—upon sprinkler activation, water is immediately discharged onto the fire. These systems are permitted in any building where the temperature is maintained at 40°F or more (Figure 9-9).

FIGURE 9-8 Riser, floor control valves and test connection for a wet-pipe automatic sprinkler system

Dry-pipe automatic sprinkler systems are used in any environment where the temperature is less than 40°F, including loading docks, attics and refrigerated storage warehouses (Figure 9-10). The piping above the dry-pipe alarm valve is charged with compressed air or nitrogen. The air or nitrogen forces the alarm valve closed so water cannot enter and freeze inside the pipe. When a sprinkler activates, it exhausts air through its discharge orifice, which lowers the pressure inside the pipe, allowing the alarm valve to open and begin flowing water toward the open sprinkler. Depending on the occupancy being

Types of Automatic Sprinkler Systems 121

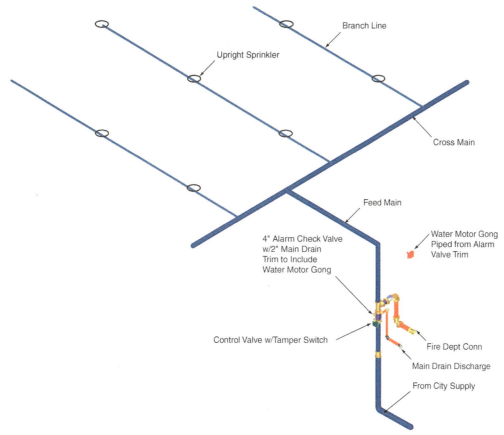

FIGURE 9-9 Wet-pipe automatic sprinkler system *(Courtesy of MFP Fire Protection Design, Gilbert, AZ)*

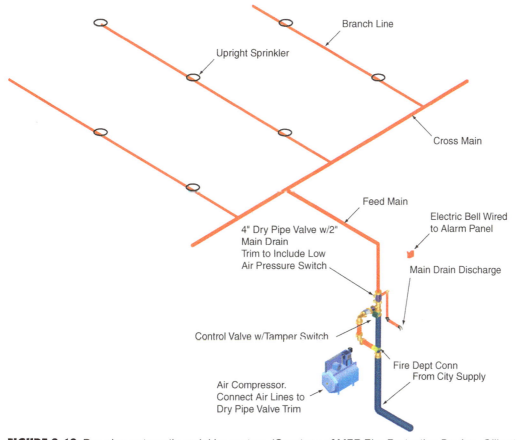

FIGURE 9-10 Dry-pipe automatic sprinkler system *(Courtesy of MFP Fire Protection Design, Gilbert, AZ)*

protected, NFPA 13 requires the discharge of water from the open sprinkler within 15 to 60 seconds of operation. On many dry-pipe sprinkler systems, a device called an accelerator, exhauster or quick-open device is installed to help accelerate the release of the air or nitrogen when a fire sprinkler operates.

A pre-action automatic sprinkler system utilizes a supplemental fire detection system located in the same area as the sprinklers (Figure 9-11). Depending on the design, the piping may be charged with compressed air. The fire detection system activates an alarm and opens the deluge valve, allowing the pipes to be filled with water. Such a design is known as a single-interlock pre-action system because only one action is necessary to allow water into the piping. A pre-action sprinkler system that only admits water into the pipe after activation of both a sprinkler and the fire detection system is termed a double-interlock system, because two actions ([1] sprinkler operation and [2] fire detection system activation) are necessary to allow water to enter the piping. NFPA 13 does not allow more than 1,000 sprinklers to be controlled by one single-interlock alarm valve. Single- and double-interlock pre-action sprinklers must discharge water within 60 seconds of the sprinkler's operation.

Deluge automatic sprinkler systems are similar to pre-action sprinklers in that they are also activated by a fire detection system located in the same area as the sprinklers (Figure 9-12). However, the sprinklers in this fire protection system are open—the heat detection element and the water seal are removed. When a deluge sprinkler system operates, water flows from all of the sprinklers simultaneously. Because all of the sprinklers flow water upon activation, the design of these systems is limited to the available water supply. Deluge automatic sprinkler systems are generally utilized when protecting very challenging goods or materials such as acetylene gas cylinder transfilling plants or storage of certain flammable liquids in plastic packaging.

> **Code Essentials**
>
> NFPA 13 sets forth design requirements for four different types of automatic sprinkler systems:
> - Wet-pipe
> - Dry-pipe
> - Pre-action
> - Deluge

OCCUPANCIES REQUIRING AUTOMATIC SPRINKLER PROTECTION

Code requirements for automatic sprinklers can be found among the various occupancy classes. Buildings housing a Group H-5, I or R fire area require the installation of automatic sprinkler protection throughout the building because of the life safety and fire protection risks. In the case of Group I occupancies (hospitals, nursing homes

Occupancies Requiring Automatic Sprinkler Protection 123

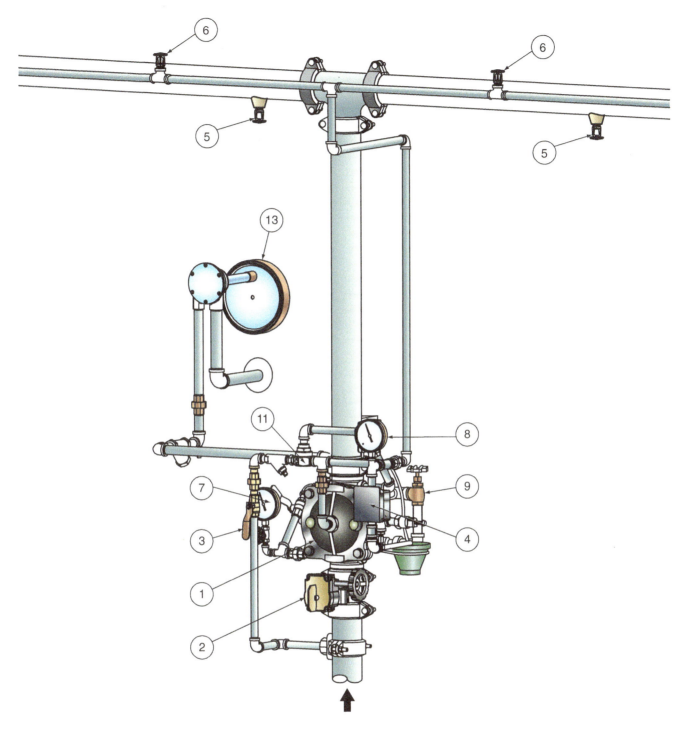

1 - Model DV-5 Deluge Valve
2 - Main Control Valve (N.O.)
3 - Diaphragm Chamber Supply Control Valve (N.O.)
4 - Local Manual Control Station
5 - Open Nozzles or Sprinkles
6 - Wet Pilot Line Sprinklers (Fire Detection)
7 - Water Supply Pressure Gauge
8 - Diaphragm Chamber Pressure Gauge
9 - System Drain Valve (N.C.)
10 - Main Drain Valve (N.C.) (Shown at Rear of Valve)
11 - Diaphragm Chamber Automatic Shut-Off Valve
12 - Waterflow Pressure Alarm Switch (Shown at Rear of Valve)
13 - Water Motor Alarm (Optional)

Note:
N.C. = normally closed
N.O. = normally open

FIGURE 9-11 Pre-action automatic sprinkler system *(Courtesy of TYCO Fire Suppression and Building Products, Lansdale, PA)*

124 Chapter 9 Automatic Sprinkler Systems

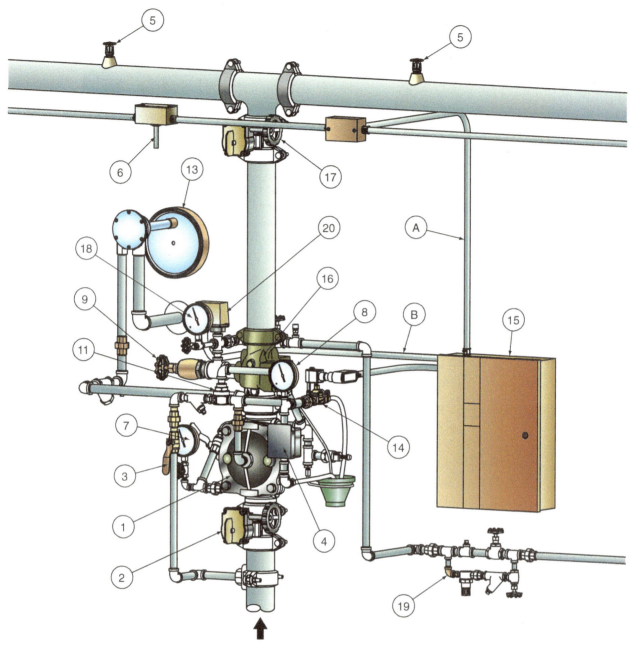

1 - Model DV-5 Deluge Valve
2 - Main Control Valve (N.O.)
3 - Diaphragm Chamber Supply Control Valve (N.O.)
4 - Local Manual Control Station
5 - Automatic Sprinklers
6 - Heat Detectors, Smoke Detectors, etc. (Fire Detection)
7 - Water Supply Pressure Gauge
8 - Diaphragm Chamber Pressure Gauge
9 - System Drain Valve (N.C.)
10 - Main Drain Valve (N.C.) (Shown at Rear of Valve)
11 - Diaphragm Chamber Automatic Shut-Off Valve
12 - Waterflow Pressure Alarm Switch (Shown at Rear of Valve)
13 - Water Motor Alarm (Optional)
14 - Solenoid Valve
15 - Cross-Zone Deluge Valve Releasing Panel
16 - Riser Check Valve
17 - System Shut-Off Valve (N.O.)
18 - Air Pressure Gauge
19 - Automatic Air/Nitrogen Supply
20 - Low Pressure Alarm Switch
A - Fire Detection Initiating Circuit (Zone 1)
B - Low Pressure Alarm Initiating Circuit (Zone 2)

Note:
N.C. = normally closed
N.O. = normally open

FIGURE 9-12 Deluge automatic sprinkler system *(Courtesy of Tyco Fire Suppression and Building Products, Lansdale PA)*

and penitentiaries), the occupants are not capable of self-rescue. Two exceptions are allowed for Group I-4 day care facilities:

1. In a Group I-4 day care located on the level of exit discharge and each care room has an exit door directly to the exterior, the fire sprinkler system is not required.
2. In a Group I-4 day care located on a story other than the level of exit discharge, the automatic sprinkler system can be installed just on that floor and all floors below, rather than throughout the entire building. [Ref. 903.2.6]

For Group R occupancies, the concern is the ability of the person to be awakened and perform rescue for themselves and any other individual in the building. Buildings housing Group H-5 occupancies (semiconductor fabrication facilities) require automatic sprinkler protection throughout because of the amount and variety of hazardous materials. [Ref. 903.2.5.2, 903.2.8]

The IFC establishes thresholds for fire areas in a building in order for the fire official to determine where an automatic fire sprinkler system is required. Once the determination is made that fire sprinklers are to be installed, the code does not always require the entire building to be equipped with fire sprinklers. Depending on the occupancy classification, the code requires fire sprinklers to be installed in the fire area, the occupancy, the entire building or just the floor. When the code requires that the floor is protected with fire sprinklers, then floor levels between that floor and the levels of exit discharge must also be protected with fire sprinklers. Table 9-2 identifies the requirements for automatic sprinkler system installation based on occupancy classification.

TABLE 9-2 Portion of building required to be equipped with fire sprinklers

Occupancy classification	Throughout the fire area	Throughout the occupancy	Throughout the entire floor and all floors to levels of exit discharge	Throughout the entire building
A-1, A-2, A-3, A-4			X	
Group A on a rooftop			X[a]	
A-5	b			
B Ambulatory Care Facility			X[c]	
E	X			
F-1 Woodworking Facility	X			
F-1, all other uses				X
H-1				X
H-2, H-3, H-4		X		d
H-5, I, M, R, S				X
S-2 Enclosed Parking Garage		X		

a. Fire sprinklers are not required on the rooftop unless a ceiling is present.
b. Fire sprinklers are required in concession stands, retail areas, press boxes and other accessory use areas > 1,000 ft^2
c. In addition to fire sprinklers between the floor and the level of exit discharge, sprinklers are required on all floors below the ambulatory care facility.
d. Group H occupancies with >100 pounds of pyroxylin plastics must sprinkler the entire building. [Ref. 903.2]

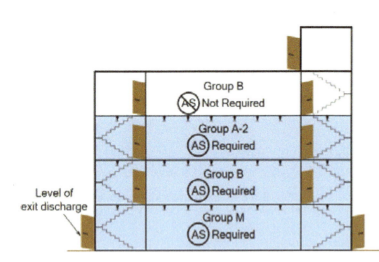

FIGURE 9-13 Automatic sprinkler system requirements for a mixed occupancy Group A-1/B/M occupancy

Code Essentials

The IFC requirements for an automatic sprinkler system are based on the building's or space's occupancy classification, its fire area, the occupant load or a combination of factors. The requirements for these systems are also based on the hazards of certain materials. •

In assembly occupancies, the requirements for automatic sprinkler protection are based on the occupancy's fire area or occupant load. In Group A-1, A-3 and A-4 occupancies, automatic sprinkler protection is required when the fire area is more than 12,000 square feet or the occupant load is 300 or more persons (Figure 9-13). In Group A-2 occupancies, the requirement for an automatic sprinkler system is more restrictive when compared to other Group A occupancies as the fire area threshold is reduced to 5,000 square feet and the occupant load threshold is reduced from 300 to 100 or more. This lower occupant load is prescribed because alcoholic beverages are served as part of the business and the concern is the ability of impaired individuals to recognize a fire and rapidly egress. In all these occupancies, sprinkler protection is required in the entire floor housing the Group A occupancy fire area. If the Group A occupancy is located on a floor that is not the level of exit discharge, sprinkler protection must be extended to each floor between the Group A occupancy and all levels of exit discharge and include the level of exit discharge. A new trend is to utilize the roof as an assembly area. Rooftops can offer an outdoor setting with a nice view, and these locations are being used for receptions, lounge areas and bars. If this assembly use were inside the building, the code would require an automatic sprinkler system be installed on the floor and all floors down to the levels of exit discharge. Since this Group A occupancy is located on the roof, providing fire sprinklers on the roof is not practical, but the installation of an automatic fire sprinkler system on all floors down to the levels of exit discharge is still required. **[Ref. 903.2.1.1, 903.2.1.2, 903.2.1.3, 903.2.1.4, 903.2.1.6]**

Where multiple fire areas are constructed in a single building, the fire areas are reviewed independently with some exceptions. In Groups F-1, M and S occupancies the fire areas are aggregated to determine the total area. In Group A occupancies, the fire areas are only aggregated if they share an exit or exit access components. Consider a strip mall with multiple restaurants each classified as Group A-2. Each restaurant is provided with exits directly to the outside. In this case, since the egress paths do not mix, each restaurant is considered separately when they are located in different fire areas. When the building configuration is modified, and the exit paths are shared, then the aggregate occupant is used to determine the fire sprinkler requirement with a threshold of 300 or more (Figure 9-14).

In Group B occupancies, automatic sprinkler protection is required only in buildings housing ambulatory care facilities or in a building that has an occupant load of 30 or more that is located more than 55 feet above the lowest level of fire department vehicle access (Figure 9-15). The IFC exempts ambulatory care facilities located on the level of exit discharge with no more than three care recipients incapable of self-preservation. Open parking garages are also exempted from these requirements, but they have their own criteria in IBC Section 406.

When a Group B building has a floor level more than 55 feet in height, the entire building must be protected with an automatic sprinkler system. However, when an automatic sprinkler system is installed as required for an ambulatory care facility, the sprinkler system is required

- on the entire floor containing the ambulatory care facility
- on all floors between that floor and the nearest level of exit discharge
- on the level of exit discharge
- on all floors below that floor. [Ref. 903.2.2, 903.2.11.3]

While each occupancy classification has its own specific requirements based on specific hazards that may be located within them, Group F-1, M and S-1 occupancies require automatic sprinkler protection when any of the following occur:

1. the fire area of any one of these occupancies exceeds 12,000 square feet;
2. any of these occupancies is located more than three stories above the grade plane;
3. the combined area of any one of these occupancies on floors of a building, including mezzanines, exceeds 24,000 square feet;
4. the fire area exceeds 2,500 square feet and houses manufacturing or storage of upholstered furniture or mattresses in Group F-1 or S-1; or
5. the fire area exceeds 5,000 square feet and houses storage or display of upholstered furniture or mattresses in Group M.

These occupancies are treated as having similar fuel packages and fire loads unless a unique hazard is introduced within them (Figure 9-16). [Ref. 903.2.4, 903.2.7, 903.2.9]

Group M and S occupancies often have products that are stored or displayed in a high-piled stor-

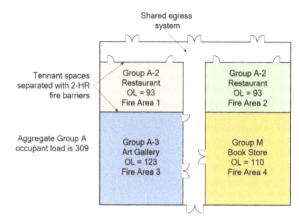

FIGURE 9-14 Each Group A fire area is below the threshold for sprinklers, but they share common exit and exit access paths, therefore the aggregate occupant is evaluated to determine automatic sprinkler system requirements.

FIGURE 9-15 This Group B occupancy requires automatic sprinkler protection because it has an occupant load of more than 30 on a floor more than 55 feet above the lowest level of fire department vehicle access.

FIGURE 9-16 Group M occupancies with a fire area over 12,000 square feet require an automatic sprinkler system.

FIGURE 9-17 Group M occupancies with combustible storage over 12 feet in height require an automatic sprinkler system when the storage area exceeds 12,000 square feet.

FIGURE 9-18 A Group H-3 compressed gas packaging plant

age configuration. With some exceptions, when combustible products exceed 12 feet in height, they are treated as high-piled combustible storage (Figure 9-17). The high-piled combustible storage results in other automatic sprinkler system requirements which are in addition to, but distinct from, those requirements above. **[Ref. 903.2.7.1]** (Additional information about high-piled combustible storage can be found in Chapter 14 of this text.)

Buildings are assigned a Group H occupancy classification whenever they store or use hazardous materials in excess of the limits allowed in IFC Chapter 50. Quantity limits, termed "maximum allowable quantity per control area" (MAQ), are established in IFC Chapter 50. (Additional information about MAQs and how they are applied can be found in Chapter 16 of this text.) The Group H occupancy requirements in the IFC and IBC are not based on the occupant load or fire area; they are dependent on the classification of the hazardous materials and the amount stored or used inside a building. When a building is constructed in accordance with the IBC Group H-1, H-2, H-3 or H-4 occupancy requirements, automatic sprinkler protection is required throughout the occupancy (Figure 9-18). In the case of Group H-1, the "occupancy" is the entire building since this occupancy is not allowed to be in a mixed-occupancy building. The hazard in a Group H-1 occupancy mandates that these occupancies are detached and separate buildings with no other uses. Buildings classified as Group H-5 require sprinkler protection throughout the building. The design of the automatic sprinkler system must comply with the requirements in IFC Chapter 50, and for many hazardous materials its design will require special fire protection considerations. **[Ref. 903.2.5]**

FIRE SPRINKLERS "THROUGHOUT" AND EXEMPT LOCATIONS

The code contains several phrases to express the requirement to install an automatic sprinkler system "throughout" the building. In some situations, the code will state "an automatic sprinkler system installed throughout the building," "the building shall be equipped throughout with an automatic sprinkler system," "an automatic sprinkler system shall be provided throughout" or "automatic sprinkler system shall be installed throughout the entire building." The code official needs to understand the meaning of the concept "throughout" the building—sprinklers are to be installed throughout the building in all of the locations required by the code and the referenced standard. As indicated earlier in this chapter, the IFC recognizes and references three standards for the installation of an automatic sprinkler system: NFPA 13, NFPA 13R and NFPA 13D. Each of these standards specifies where fire sprinklers are to be located within the building. Each of these standards also allows certain areas of the building to be unsprinklered. For example, NFPA 13D does not require sprinklers to be installed in some attic spaces, small bathrooms and closets, exterior balconies or attached garages; NFPA 13R does not require sprinklers to be installed in some attic spaces, small bathrooms and closets or some exterior balconies and decks; NFPA 13 does not require sprinklers in soffits, eaves and overhangs less than 4 feet wide, concealed spaces with limited access and no storage, some pipe chases and concealed spaces providing less than 6 inches of clearance. The fact is, none of the standards mandate that sprinklers are to be installed in all locations of the building. There are locations within a building that can be left unsprinklered, and the building will still meet all the requirements of the standard. The intent of the code is to have fire sprinklers installed "throughout" the building as required by the code and the standard. When fire sprinklers are installed in all the locations that the code and the standard requires, the building is considered to be "sprinklered throughout." This would even apply to a single-family dwelling with sprinklers installed in accordance with NFPA 13D. If sprinklers are installed in all required locations, the building is sprinklered throughout. [Ref. 903.3.1.1, 903.3.1.2, 903.3.1.3]

The concept of "sprinklered throughout" calls for sprinklers where required by the installation standard and by the IFC. The IFC contains some sprinkler installation requirements and based on Section 102.7.1, the requirements in the code supersede the requirements in the standard. For example, Section 903.3.1.1.2 allows the elimination of sprinklers in bathrooms of Group R occupancies provided the bathroom does not exceed 55 square feet, is within a dwelling unit or sleeping unit, and the walls are of limited combustible materials with a 15-minute thermal barrier. Gypsum board,

$1/_2$-inch thick, can provide this rating. Section 903.3.1.2.1 requires that sprinklers are provided on exterior decks and balconies when the building is of Type V construction even though NFPA 13R does not require such installation. The IFC provides additional criteria for sprinkler installation or deletion based on specific hazards and uses.

The IFC further specifies areas within a building that are exempt from sprinkler installation. Section 903.3.1.1.1 lists six locations in buildings that can be left unsprinklered if the room is equipped with an automatic fire detection system.

1. Sprinklers can be left out of rooms where they can create a life or fire hazard if they were to operate.
2. Sprinklers can be left out of rooms where they are considered undesirable by the code official based on the nature of the room contents.
3. Rooms where the construction is noncombustible and the contents are noncombustible can be unsprinklered.
4. Generator and transformer rooms can be unsprinklered if they are separated by fire-resistance-rated construction of at least 2 hours.
5. Rooms and spaces containing elevator machinery or elevator controls for fire service access elevators do not require sprinklers.
6. Rooms and spaces containing elevator machinery or elevator controls for occupant evacuation elevators do not require sprinklers.

In these six instances, the IFC allows the sprinklers to be omitted, but the room must have a fire detection system installed in lieu of the sprinkler protection. The elimination of fire sprinklers in these situations is not recognized in the sprinker installation standards. As discussed in Chapter 1 of this book, the code requirements supersede the standard requirements even when less restrictive. Where sprinklers are left out of rooms or areas as allowed in this section, the building is still considered sprinklered throughout as long as all other areas of the building are protected as required by the appropriate standard. [Ref. 903.3.1.1.1]

FIRE DEPARTMENT CONNECTION

A fire department connection (FDC) is required for most NFPA 13 and 13R automatic sprinkler systems and standpipe systems. An FDC is not required for automatic sprinkler systems protecting one- and two-family dwellings and townhouses. Fire apparatus can connect supply hoses to the FDC to pump additional water into sprinkler or standpipe systems. The piping arrangement between the FDC and the sprinkler riser depends on the type of automatic sprinkler system. [Ref. 903.3.7, 912.1]

The location of and the fire hose threads installed on an FDC must be approved by the fire code official. The FDC's placement must not obstruct access to the protected building for other responding apparatus. The connection is located on the street side of buildings, fully visible from the street and it must be easily recognized from the fire department vehicle access roadway (Figure 9-19). [Ref. 912.2]

To ensure that the water supply for the sprinkler or standpipe system is supported by fire apparatus, it is important that the FDC's location be identified. For existing buildings, the IFC authorizes the code official to require the installation of additional signs to help identify the FDC location (Figure 9-20). In many cases, the IFC and IBC only require automatic sprinkler protection for the fire area or a portion of the building. In such cases, a sign is required at the FDC to indicate the portion of the building served. In Group R-2 apartment complexes with multiple buildings, it is common to have a single FDC supplying multiple buildings, because it is more economical to provide separate water supply connections to serve a limited number of buildings. In these cases, the buildings served by the FDC should be identified. Ref. 912.2.2, 912.4]

FIGURE 9-19 Fire department connections are to be located on the street side of buildings, visible from the street and provided with ready access.

FIGURE 9-20 A direction sign can be required at existing buildings where the FDC is not readily visible from the street.

CHAPTER 10
Fire Alarm and Detection Systems

Smoke alarms for occupant notification in one- and two-family dwellings and townhouses, local fire alarm and detection systems that notify building occupants, and automatic sprinkler systems connected to a supervising central station are examples of fire alarm and detection systems regulated by the *International Fire Code* (IFC). Fire alarm and detection systems provide early warning of a fire by detecting fire's products of combustion such as smoke, heat or a visible flame. Pre-action and deluge sprinkler systems and most alternative-agent fire-extinguishing systems are activated by an automatic fire detection system connected to a fire alarm control unit. These systems are specified by the IFC in occupancies that present high life-safety risks or buildings where a large population is present and require a reliable means of early occupant notification and communication.

DESIGN AND INSTALLATION STANDARDS

Requirements for the design, construction and maintenance of fire alarm and detection systems are set forth in NFPA 72, *National Fire Alarm and Signaling Code*. These systems utilize electricity, so the wiring of these systems must also comply with NFPA 70, *National Electrical Code®*. NFPA 72 stipulates the requirements for devices to initiate a fire alarm signal, transmit a signal to and from the fire alarm control unit and notify the occupants both visually and audibly. It also contains requirements for the performance, reliability and survivability of the fire alarm and detection system (Figure 10-1).

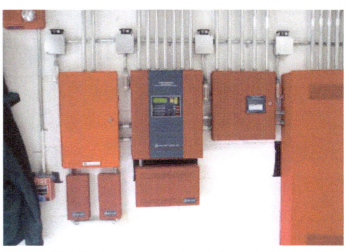

FIGURE 10-1 The fire alarm control unit is the control center for the fire alarm system. Signals are received from initiating devices, and signals are sent out to operate notification devices.

A fire alarm and detection system is designed to perform several functions:

- Notifying of a fire alarm in a building
- Monitoring and notifying of supervisory and trouble signals
- Alerting the occupants
- Summoning aid
- Controlling fire safety functions. [Ref. 202]

In the context of NFPA 72, notification is an audible, visual or text signal, message or display. Requirements for when notification devices are required are based on the building occupancy classification and occupant load. Audible alarm notification appliances must emit a distinctive signal that cannot be used for any purpose other than that of a fire alarm (Figure 10-2). In any fire alarm and detection system, occupant notification takes precedence over any supervisory signals. [Ref. 907.5.2]

Supervisory signals indicate the need for action in connection of supervising guard tours, fire suppression systems or maintenance. Supervisory signals include the closing of a main water supply control valve to a fire pump or standpipe system, low air pressure on a dry-pipe automatic sprinkler system, or any condition that takes a fire protection system off-line. A supervisory signal must be distinctive from other signals that are received and must be visually annunciated at a fire alarm control unit. A supervisory signal warrants implementation of the building's fire protection impairment program (Figure 10-3). [Ref. 202]

FIGURE 10-2 Audible and visible occupant notification is provided by this device.

FIGURE 10-3 Valve tamper switches electrically supervise the wall post indicator valves for automatic sprinkler systems and send a tamper signal to the fire alarm control unit.

FIGURE 10-4 Motorized fire damper in a fire-resistance-rated wall controlled by the fire alarm control unit.

FIGURE 10-5 Magnetic door hold-open devices serve a fire safety function and operate upon a signal from the fire alarm control unit.

Fire alarm and detection systems also can perform fire safety functions. A fire safety function is any feature that improves occupant life safety or controls the spread of fire and smoke. Fire safety functions include activation of motorized controls to close fire and smoke dampers (Figure 10-4) Fire safety functions also include the release of magnetic devices that hold open protection such as fire door assemblies that protect openings in fire walls or barriers. Upon activation of the fire alarm and detection system, the magnet is de-energized, causing the fire door to close and protect the opening in the wall (Figure 10-5). Activation of devices performing fire safety functions in buildings that are required to be equipped with a fire alarm and detection system must transmit a distinct audible and visual supervisory signal to a constantly attended location or activate the occupant notification devices. [Ref. 907.3]

Duct smoke detectors perform fire safety functions (Figure 10-6). These specialized smoke detectors are required by Section 606.2 of the *International Mechanical Code* when the airflow rate in a return airflow duct or plenum system is more than 2,000 cubic feet per minute. Duct smoke detectors constantly monitor the air for the presence of smoke in the return air duct or plenum. The return air ductwork carries air from the occupied spaces to the heating, ventilating and air conditioning units. The detector is listed for use in mechani-

FIGURE 10-6 Duct smoke detector
(Courtesy of Air Products and Controls, Pontiac, MI)

cal ventilation systems based on the duct diameter or width, the airflow velocity and the air temperature and humidity range of the air-handling system. When a fire occurs within the building, the return air system will pick up the smoke and, if the air handling equipment is not shut down, will distribute the smoke back into the building through all of the supply air ductwork. Activation of a duct smoke detector generally shuts down the air-handling system, unless it serves a building smoke control system. In this case, the building's mechanical system is switched to a smoke control mode. [Ref. 907.3.1]

All IFC required fire alarm systems must be monitored by an approved supervising station (Figure 10-7). A supervising station is a facility that receives fire alarm and supervisory signals and transmits them to the fire department. The supervising station is constantly staffed to respond to and process alarm signals. Monitoring is not required for smoke alarms or dwelling fire sprinkler systems in one- and two-family dwellings, nor is monitoring required for smoke detectors in Group I-3 occupancies. The design, installation and maintenance of the monitoring circuits must comply with NFPA 72. [Ref. 907.6.6]

NFPA 72 requires certified and competent fire protection system designers and installers for the design, installation, testing and maintenance of fire alarm and detection systems. The IFC requires the system be tested upon completion of the installation in accordance with NFPA 72. These test results are documented on a record of completion that is provided to the fire code official. [Ref. 907.8, 907.8.5]

During the life of the occupancy, the owner is responsible for the continued maintenance of the fire alarm and detection system. The testing frequencies for various types of detectors, the fire alarm control unit, audible and visible signal strength and coverage and testing of devices performing fire safety are specified in NFPA 72. Within the first year of installation and every alternate year thereafter, all smoke detectors must be subjected to a sensitivity test. This test is performed to ensure smoke detectors are operating within their sensitivity range and listing (Figure 10-8). [Ref. 907.8.3]

FIGURE 10-7 This digital alarm communications transmitter sends fire alarm, supervisory and trouble signals to the supervising station. *(Courtesy of Honeywell Security and Communications)*

Code Essentials

NFPA 72 is adopted by the IFC and governs the design, inspection, testing and maintenance of fire alarm and detection systems. These systems commonly consist of
- Initiating devices
- Notification appliances
- Fire alarm control unit
- Components that serve a fire safety function

FIGURE 10-8 Functional test of a duct smoke detector *(Courtesy of Air Products and Controls, Pontiac, MI)*

FUNDAMENTAL COMPONENTS

All fire alarm and detection systems required by the IFC have four fundamental components:

- Fire alarm control unit
- Initiating devices
- Occupant notification devices
- A primary and secondary electrical power supply

The fundamental components are required for all protected premises, as defined in NFPA 72. These components are not found in single- or multiple-station smoke alarms required in Group R or certain I occupancies. All components of a fire alarm system must be compatible with the fire alarm control unit, approved by the fire code official and listed by a nationally recognized testing laboratory. [Ref. 907.1.3]

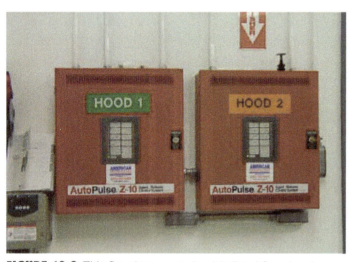

FIGURE 10-9 This fire alarm control unit is listed for releasing service of an alternative automatic fire-extinguishing system.

A fire alarm control unit is the component that receives input from automatic and manual initiating devices. The control unit may be capable of supplying electrical power to detection devices, notification appliances and transponders that transmit signals to a supervising station. A fire alarm control unit can serve part or all of a building. For pre-action and deluge sprinkler systems and many alternative fire-extinguishing systems, the control unit must be listed as a releasing service fire alarm control unit, because it controls the activation of a fire-extinguishing system (Figure 10-9). A fire alarm control unit must actuate notification devices and fire safety functions and annunciate the location of the initiating device within 10 seconds after its activation. The fire alarm control unit is used by fire fighters to identify the location of the device that initiated the fire alarm signal and assist in locating the fire in large complexes. Unless the fire alarm control unit is located in a continuously occupied area, the IFC requires that all fire alarm control units be protected by an approved automatic fire or smoke detection system. [Ref. 202, 907.4.1]

Fire alarm control units are required by NFPA 72 to have a primary and secondary power supply. The primary power supply is commonly an electric utility serving the normal building power. The source of the power is required to be supplied by a dedicated fire alarm branch circuit. The branch circuit must be protected from mechanical impact and identified at the circuit breaker so that power is not inadvertently turned off.

The secondary power supply can be any source allowed by NFPA 72, and it is commonly accomplished using storage batteries. The secondary power supply must be capable of supplying power for at least 24 hours and, in the event of system activation, be able to operate all notification appliances for at least 5 minutes. If the fire alarm control unit is serving an emergency voice/alarm communication system, NFPA 72 requires that the secondary power source be capable of operating for a minimum of 15 minutes. [Ref. 907.6.2]

Initiating devices are connected to the fire alarm control unit. Initiating devices are components that originate a change-of-state condition, and include photoelectric or ionization smoke detectors, heat detectors, a manual fire alarm box, a sprinkler water flow or pressure switch or a supervisory signal (Figure 10-10). Initiating devices are available that can identify their specific location and function at the fire alarm control unit. These types of devices are referred to as addressable devices; each device has a specific address in the fire alarm system, and the fire alarm control unit can determine which specific device operates. Addressable systems are required for new fire alarm installations unless the building is only one story with less than 22,500 square feet and the fire alarm initiating devices are limited to water-flow devices, manual fire alarm boxes and no more than 10 additional initiating devices. NFPA 72 requires initiating devices to be located so they are accessible for maintenance and testing and in all areas and locations prescribed by the IFC. If they can be subject to mechanical damage, NFPA 72 requires the initiating device be protected with a mechanical guard. [Ref. 907.4, 907.6.3]

One initiating device that is accessible to all building occupants is a manual fire alarm box (Figure 10-11). A manual fire alarm box is a device used by the public to initiate a fire alarm signal. Manual fire alarm boxes are normally required by the IFC in any occupancy where a fire alarm and detection system is required. If the building is protected by an automatic sprinkler system, manual fire alarm boxes throughout the building are not required in many occupancies, provided the occupant notification devices activate when the automatic sprinkler system operates and at least one manual fire alarm box is provided in a location approved by the fire code official. [Ref. 907.2]

Manual fire alarm boxes must be located within 5 feet of each exit on each floor. In buildings that are not equipped with an automatic sprinkler system, additional manual fire alarm boxes are required so that the travel distance from any location in the building to the fire alarm box is limited to 200 feet. Boxes must be installed 42 to 48 inches above the finished floor and must be red in color. If the fire alarm system is not monitored by a supervising station, a sign is required above each box stating that the fire department must be notified when the fire alarm box is activated. [Ref. 907.4.2]

Many times manual fire alarm boxes are located in areas where they can be struck or hit during normal operations or use of the build-

FIGURE 10-10 A sprinkler water-flow switch and valve-tamper switch are examples of initiating devices. The control valve tamper switch is a supervisory device.

FIGURE 10-11 Manual fire alarm box within 5 feet of exit

FIGURE 10-12 Protective cover over a manual fire alarm box

FIGURE 10-13 Audible and visual occupant notification appliance

ing. To avoid accidental activations or damage, the fire code official can require protective covers to be placed over the manual fire alarm boxes. The covers must be listed and must be transparent so the occupants can see the manual fire alarm box (Figure 10-12). The box cannot be an obstruction to the egress path and can only protrude 4 inches into the required clear width. **[Ref. 907.4.2.5]**

Occupant notification appliances are required for many of the fire alarm systems prescribed by the IFC. Notification appliances are grouped into notification zones, which are areas of a building where all of the notification appliances simultaneously operate when an alarm signal is received and processed by the fire alarm control unit. Notification appliances are designed to deliver audible, visible, tactile or a combination of these signals (Figure 10-13). **[Ref. 202]**

Occupant notification appliances activate when an initiating device operates. The audible alarm signal must be distinct and cannot be used for any other purpose. Audible notification appliances must be located so they can be heard above the sound level in the building. The minimum sound level for the notification appliances must be at least 15 decibels (dBA) above the average sound level and 5 decibels (dBA) above the maximum sound level in the building during normal operation. This provides a level where the fire alarm can be noticed in noisy facilities. To protect the occupants' hearing, the notification devices cannot exceed 110 dBA, so when the average ambient level exceeds 95 dBA, visible alarm notification appliances shall be provided for occupant notification. **[Ref. 907.5.2.1]**

Group A occupancies with an occupant load of 1,000 or more, Group E with an occupant load of 100 or more, special amusement buildings, atriums in Group A, E or M, malls over 50,000 square feet, high-rise buildings and deep underground buildings require an emergency voice/alarm communication system. This occupant notification system is activated in the same manner as for audible and visible occupant notification appliances. However, this system requires the installation of a speaker network that delivers recorded or live verbal messages to the building occupants. The system will have pre-recorded messages, with different messages applicable to different situations based on the initiation device activated. The system also will be capable of delivering live voice messages during the emergency. This allows fire fighters to direct occupants to perform either a complete, partial or staged evacuation (Figure 10-14). In high-rise buildings, the emergency voice/alarm communication system must transmit signals to the floor where the alarm-initiating device operated and the floor above and below this level. Individual paging zones are also required for each floor, within interior exit stairways, at elevator groups and areas of refuge. The system must be designed so it can be manually controlled by emergency responders and be connected to an emergency power source. [Ref. 907.2.1.1, 907.2.3, 907.2.11.3, 907.2.12, 907.2.13, 907.2.18, 907.5.2.2]

OCCUPANCIES REQUIRING FIRE ALARM AND DETECTION SYSTEMS

FIGURE 10-14 An emergency voice/alarm communication system integrated into a fire alarm control unit *(Courtesy of Siemens Building Technologies, Florham Park, NJ)*

The IFC sets forth requirements for fire alarm systems in new and existing buildings. The IFC requires the installation of automatic fire detection in occupancies where early detection and warning of an incipient fire is critical to occupant life safety. In certain existing buildings, the IFC requires the retroactive installation of these systems if they are not currently protected by a fire alarm system. Requirements for occupant notification depend on the occupancy classification and occupant load of the building. The IFC has specific requirements for these systems in certain buildings such as covered malls, underground buildings, stadiums and high-rise buildings. They are not required in Group S and U occupancies. [Ref. 907.2]

In most occupancies, the IFC allows the elimination of manual fire alarm boxes when the building is protected throughout by an automatic sprinkler system designed in accordance with NFPA 13 and the occupant notification appliances activate upon sprinkler water flow. However, at least one manual fire alarm box is required at an approved location. This manual fire alarm box is not required if the fire alarm system is dedicated to elevator recall or supervisory service, or in Group R-2 occupancies unless it is required by the fire code official. [Ref. 907.2]

Assembly occupancies require a manual fire alarm system that activates the occupant notification system when the occupant load is 300 or more, or if the occupant load is more than 100 persons above or below the level of exit discharge (Figure 10-15). If the building has an occupant load of 1,000 or more, occupant notification must be performed using an emergency voice/alarm communication system. [Ref. 907.2.1]

FIGURE 10-15 This Group A-4 occupancy with an occupant load of 1,000 or more requires a manual fire alarm system with an emergency voice/alarm communication system.

With the exception of ambulatory care facilities, Group B occupancies only require a manual fire alarm system when the combined occupant load on all floors is 500 or more or if the occupant load is more than 100 persons above or below the level of exit discharge. Group B ambulatory care facilities require a smoke detection system throughout the occupancy and in all common areas leading to the exits. If the ambulatory care facility is protected with an automatic sprinkler system, the smoke detectors can be eliminated, but the sprinkler system must activate the occupant notification appliances. [Ref. 907.2.2, 907.2.2.1]

All Group E occupancies with an occupant load of more than 50 require a manual fire alarm system that activates the occupant notification system. When these buildings are not protected by an automatic sprinkler system, manual fire alarm boxes can be eliminated when the interior corridors are protected by smoke detectors; auditoriums, cafeterias, gymnasiums, shops and laboratories are protected by an approved means of fire detection; and the occupant notification system can be activated at a central point. When the occupant load exceeds 100, the fire alarm system must be an emergency voice/alarm communication system (Figure 10-16). [Ref. 907.2.3]

FIGURE 10-16 This Group E occupancy with an occupant load of more than 100 requires a manual fire alarm system with an emergency voice/alarm communication system.

FIGURE 10-17 Visible notification device used as an element of private mode signaling (*Courtesy of Cooper Notification Inc.*)

Group I occupancies represent a high life-safety risk because patients may not be capable of self-preservation. In the case of correctional and detention facilities, public safety mandates the housed individuals be closely supervised. Accordingly, the IFC requires the installation of automatic smoke detection that activates the occupant notification system or in some cases only alerts the staff. Activation of any other IFC-prescribed fire protection systems also must activate the notification appliances in these occupancies. [Ref. 907.2.6]

In certain critical health care areas, it may be desirable to reduce or completely eliminate notification appliance audibility. In intensive care or surgery suites, audible notification appliances can compromise patient care. In these cases, the design professional can use private mode signaling with the approval of the fire code official and inclusion of staff evacuation responsibilities in the fire safety and evacuation plan. In private mode signaling, the sound pressure level is reduced or may be eliminated. NFPA 72 requires the installation of visible notification appliances when private mode signaling is used (Figure 10-17). [Ref. 907.2.6, Exception 2]

Assisted living and board and care facilities with more than 16 residents are classified by the *International Building Code* as a Group

Code Essentials

Group I-1, Condition 1 consists of occupancies where all persons receiving custodial care are capable of responding to an emergency situation and complete building evacuation.

I-1 occupancy (Figure 10-18). In a Group I-1, Condition 1 occupancy protected with an automatic sprinkler system complying with NFPA 13, smoke detection is not required in habitable spaces, which are spaces in a building for living, sleeping, eating or cooking. Bathrooms, closets and storage or utility spaces are not habitable spaces. An automatic smoke detection system is required in corridors and waiting areas open to corridors. Single- and multiple-station smoke alarms are required in the patient sleeping area, unless the area is served by the automatic smoke detection system. [Ref. 907.2.6.1]

FIGURE 10-18 Group I-1 assisted living facility

Hospitals, intermediate care, skilled nursing homes, detoxification facilities and other Group I-2 occupancies require automatic smoke detection systems. Smoke detection is required in corridors and spaces that are permitted by the IBC to be open to corridors, such as patient visiting areas and nurse stations in Group I-2, Condition 1. Group I-2, Condition 2 occupancies require the detectors in the corridors plus other locations such as waiting rooms and psychiatric treatment areas. In Group I-2 occupancies, automatic smoke detection is not required in corridors of smoke compartments protecting patient sleeping rooms when each sleeping room is protected by a smoke detector. Activation of the sleeping room smoke detector in turn activates a visual display located outside of the patient's room in the corridor. The smoke detector's activation must initiate an audible and visual alarm at the nurse station that attends to the care of the patient (Figures 10-19 and 10-20). [Ref. 907.2.6.2, Exception 1; IBC 407.2]

FIGURE 10-19 Patient sleeping room visual display *(Courtesy of West Com Nursing Call Systems, Fairfield, CA)*

A second option in Group I-2 occupancies is integrating the smoke detector into a patient sleeping room door closer (Figure 10-21). Activation of the smoke detector releases the door closer, causing the patient's room door to close and activates the occupant notification system. [Ref. 907.2.6.2, Exception 2]

FIGURE 10-20 Nurse call system with a graphic user interface *(Courtesy of West Com Nursing Call Systems, Fairfield, CA)*

Code Essentials

Group I-2, Condition 1 consists of hospitals, nursing homes, psychiatric hospitals which do not have emergency care, surgery, obstetrics or in-patient stabilization units for psychiatric care or detoxification. •

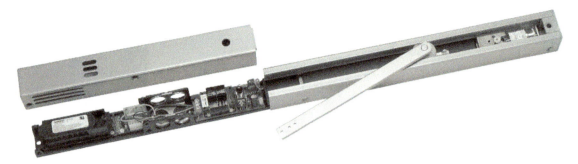

FIGURE 10-21 Door closer with an integral smoke detector *(Courtesy ASSA ABLOY, New Haven, CT)*

Similar to Group B, Group M occupancies require a manual fire alarm system when the combined occupant load on all floors is 500 or more or if the occupant load is more than 100 persons above or below the level of exit discharge. However, occupant notification in Group M can be handled differently. The code allows the initial notification to be sent to a constantly attended location during times that the building is occupied. From the constantly attended location, an emergency voice/alarm communication system can be activated and utilized with live messages. [Ref. 907.2.7]

In Group R-1 occupancies, a manual fire alarm system is required unless the building is not more than two stories and each sleeping unit is separated by 1-hour fire-resistance-rated construction and has a direct exit to the exterior. Since the occupancies are also required to be protected by an automatic sprinkler system, only one manual fire alarm box is required at an approved location. In Group R-1 occupancies, an automatic smoke detection system that activates the occupant notification system is required throughout all interior corridors that serve sleeping units. [Ref. 903.2.8, 907.2.8, 907.2.10]

FIGURE 10-22 Group R-2 apartment building

Group R-2 occupancies require a manual fire alarm system when they contain more than 16 dwelling units, dwelling units are more than two stories above the level of exit discharge or dwelling units are located below the level of exit discharge (Figure 10-22). The IFC also requires an automatic sprinkler system that complies with either NFPA 13 or NFPA 13R in these occupancies. When the building is equipped with a sprinkler system, manual fire alarm boxes are not required provided that the operation of the sprinkler water flow switch activates the occupant notification devices. In college or university buildings classified as Group R-2, smoke detection is required in common areas, interior corridors, laundry rooms and storage rooms. [Ref. 903.2.8, 907.2.9]

CARBON MONOXIDE ALARMS

Carbon monoxide is an odorless and colorless gas, and it can be deadly. Carbon monoxide is produced when the use of a solid, liquid or gaseous fuel does not result in a complete combustion process. In these instances, carbon monoxide is produced. Some examples where carbon monoxide can be produced are a gas or wood fireplace, a gas stove, heating/ventilation/air-conditioning equipment powered by fuel oil or gas, or a vehicle in the garage. A study concluded in 2015 and reported that over a 12-year period an average of 21,000 people seek medical treatment for non-fire-related carbon monoxide poisoning per year. Of those, an annual average of 2,300 are admitted to the hospital for treatment. In this 12-year period, 5,149 fatalities occurred as a result of non-fire-related carbon monoxide poisoning in the United States. The average annual fatality rate for carbon monoxide exposures is 438.[4] In the study, the most common causes of non-fire-related carbon monoxide poisoning are indoor use of fuel-fired generators, gasoline powered engines and malfunctioning heating and cooking appliances (Figure 10-23).

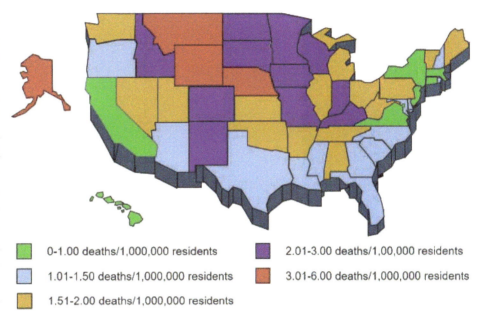

FIGURE 10-23 Unintentional carbon monoxide deaths by state, 1999 through 2012[5]. Age-adjusted rates per 100,000 resident populations.

In response to these fatalities, the IFC requires carbon monoxide alarms to be installed in Group I-1, I-2, I-4 and R occupancies and in Group E classrooms located in buildings containing fuel-burning appliances or with an attached garage (Figure 10-24). Carbon monoxide alarms have a similar design, and similar requirements, to smoke alarms. The alarms need to be connected to the building electrical system and must have a battery backup. The carbon monoxide alarms must be listed under UL 2034, Standard for Single and Multiple station Carbon Monoxide Alarms, or UL 2075, Standard

FIGURE 10-24 Combination smoke and carbon monoxide alarm

---

4. Kanta Sircar, Jacquelyn Clower, Mikyong Shin, Cathy Bailey, Michael King and Fuyuen Yip, "Carbon Monoxide Poisoning Deaths in the United States, 1999-2010," *American Journal of Emergency Medicine*, September 2015, Vol. 33, Issue 9, p. 1140-1145.
5. Ibid.

for Gas and Vapor Detectors and Sensors, but they could also be a combination carbon monoxide/smoke alarm. The typical installation is independent carbon monoxide alarms; however, a carbon monoxide detection system could be substituted, and in situations where smoke alarms and carbon monoxide alarms are required a listed combination smoke/carbon monoxide alarm can be installed. **[Ref. 915]**

GAS DETECTION SYSTEMS

The IFC deals with many flammable or toxic gases. Many of these gases require gas detection systems because of the significant hazards they present. For example, hydrogen gas detection in hydrogen fuel gas rooms, repair garages working on fuel systems of vehicles fueled by hydrogen or nonodorized liquefied natural gas (LNG), certain hazardous production material (HPM) rooms in Group H-5 occupancies, or rooms containing liquid carbon dioxide for beverage dispensing operations. There are several different sections of the code that require gas detection systems and refer to Section 916 for design and installation of the gas detection system.

Gas detection systems require an installation permit under the IFC. As part of the application for permit review, the fire code official needs to ensure that the gas detectors are appropriate for the installation. Each detector is designed to detect certain gases—the manufacturer's instructions need to be consulted and followed when selecting the proper detector (Figure 10-25).

The gas detection system must be provided with standby or emergency power, or designed to provide a signal at an approved location where normal power is lost. Gas detection systems are only to be connected to a fire alarm system when approved by the fire code official and then it must be in accordance with the fire alarm equipment manufacturer's instructions.

> **Code Essentials**
>
> A gas detection system shall activate an alarm when gas concentrations reach
> - 25 percent of the lower flammability limit (LFL), or
> - 50 percent of the value for immediately dangerous to life and health (IDLH)
>
> whichever is appropriate for the gas being detected.

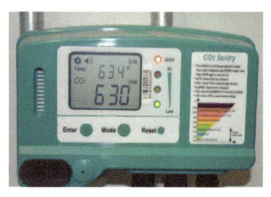

FIGURE 10-25 Carbon dioxide detection is required in rooms or areas using carbon dioxide enrichment systems.

CHAPTER 11
Means of Egress

The *International Fire Code* (IFC) sets forth requirements for the fire protection and life safety of building occupants. The greatest concern and highest safety priority are the building occupants. The IBC specifies requirements for the design of an exit, exit access and exit discharge in all buildings. The IFC prescribes detailed regulations so individuals who are capable of rescuing themselves and other building occupants can safely and expeditiously leave a building in a fire or other emergency. The requirements found in IFC Chapter 10 are termed means of egress. The IFC has requirements for means of egress in two chapters.

Chapter 10 in the IFC and *International Building Code* (IBC) are the same and are applied to any new building or occupancy. A second set of requirements in IFC Chapter 11 is applied to existing buildings. The means of egress requirements in Chapter 11 are based on the occupancy classification and adopted building code at the time of the building's construction. In some jurisdictions, this may be an earlier edition of the IBC, one of the "legacy"[6] codes or the community's own building code. If a conflict arises between the requirements in the applicable building code and IFC Chapter 11, the most restrictive requirement is applied. In cases where a building is built in a jurisdiction that did not adopt a building code at the time of construction, the means of egress system must be retrofit to meet the minimum requirements in Chapter 11. [Ref. 1104.1]

INTRODUCTION TO MEANS OF EGRESS

A building means of egress system has three distinctive and connected components:

- Exit
- Exit access
- Exit discharge [Ref. 202]

To evaluate the means of egress, the exits from a room, building or portion of a building must be determined. Any egress path leading from an exit to the public way is an exit discharge.

The exit is "that portion of a means of egress system between the exit access and the exit discharge or public way" (Figure 11-1). An "exit" is a specific component in the egress path and can consist of any of the following:

- an exterior exit door at the level of discharge
- an interior exit stairway and ramp
- an exit passageway
- an exterior exit stairway and ramp
- a horizontal exit. [Ref. 202]

Even though many building features are commonly referred to as exits, those seven components are the only items actually referred to as an exit in the code. Exits cannot be used for any purpose that interferes with its function as a means of egress component. Once an exit is required to

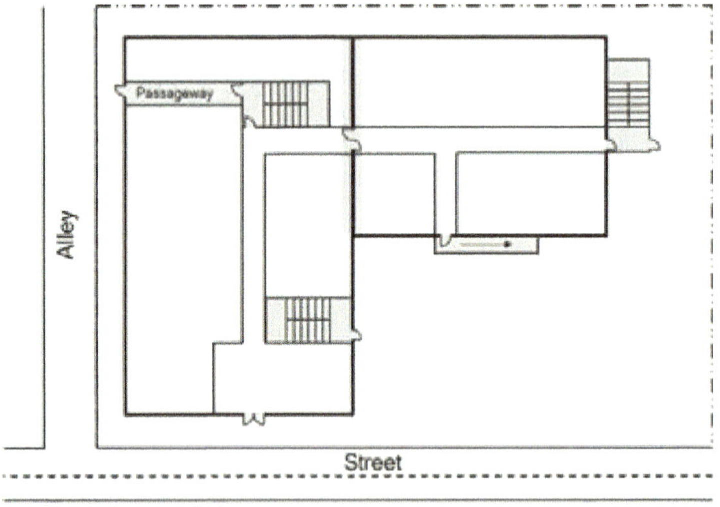

FIGURE 11-1 Exit

---

6. The International Conference of Building Officials' *Uniform Building Code*, the Southern Building Code Congress International's *Standard Building Code*, and the Building Officials and Code Administrators' *National Building Code* were consolidated in 2000 into the International Code Council *International Building Code®*.

be constructed as a fire-resistive assembly or is equipped with automatic sprinkler protection, the required level of protection cannot be reduced until it terminates at the exit discharge. The minimum number of exits required is based on the occupant load per story and the travel distance to the closest exit. Depending on occupancy classification, only one exit may be required given the building occupant load and the location of the occupancy in relation to the building's grade plane. [Ref. 202, 1022]

Exit access is "that portion of a means of egress system that leads from any occupied portion of a building or structure to an exit" (Figure 11-2). Exit access is the area of the room or space where egress commences. For the most part, any walking surface inside a building is a component in the means of egress system. The space inside a hotel room, factory, shopping mall or high-rise building is designated as either an exit access or exit. The IBC and IFC means of egress requirements ensure that any building equipment or fixtures do not impede, limit or obstruct one's ability to safely leave a building provided that the number of persons within a given space or area does not exceed the prescribed occupant load limits and travel distances to an exit. Exit access includes corridors that physically guide a person to an exit. The exit access must comply with the general provisions for all means of egress components, including ceiling height, protruding objects, continuity and walking surfaces. It also must comply with the doorway, travel distance and corridor requirements for exits and exit access. [Ref. 202, 1016]

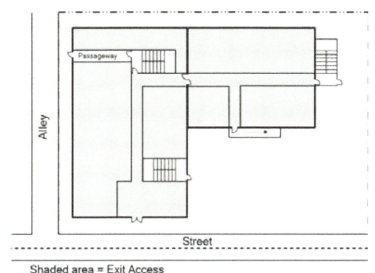

FIGURE 11-2 Exit access is the path from any point in the building to an exit

A person has not completed egress from a building until they reach a public way such as a street or alley. Exit discharge is defined as "the portion of a means of egress system between the termination of an exit and a public way" (Figure 11-3). The exit terminates where the exit discharge commences. In a multi-story building, the level of exit discharge is the floor where the occupant leaves the building or stairwell. An exit discharge must be at grade or provide a means of access grade, and cannot reenter the building. [Ref. 202, 1028]

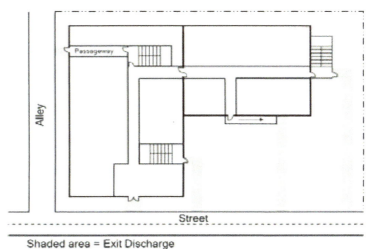

FIGURE 11-3 Exit discharge continues to the public way

> **Code Essentials**
>
> Means of egress is comprised of three interconnected components with their own distinct but complementary requirements:
> - Exit access
> - Exit
> - Exit discharge

All three components together comprise the means of egress. For example, consider people exiting from the ninth story of a building. As they start out on their exit path in a room on the ninth floor, they are in the exit access. They would continue in the exit access until they reach an exit; in this case it is an interior exit stairway. As they pass through the door into the interior exit stairway, they enter the exit. Traveling down the stairs, they are in the exit. Upon reaching the ground floor, they open the door to the outside. As they step through this door, they are now in the exit discharge. They remain in the exit discharge until they reach the public way (Figure 11-4).

Because of the size of a building or the site's topography, it may not be possible to have a direct and unobstructed path to a public way. In this case, the IBC requires an area outside of the building that can safely accommodate each occupant. A minimum area of 5 square feet is required for each person and the area must be located at least 50 feet from the building. The area must be permanently maintained and identified as a safe dispersal area (Figure 11-5). **[Ref. 1028.5]**

Every means of egress has minimum features and characteristics that must be properly designed and maintained compliant.

1. Any given egress component requires a certain minimum width that is dictated by the occupant load and minimum egress component requirements. **[Ref. 1005.1]**
2. The number of occupants is limited in a given area based on the building's use and function. The number of persons allowed in the given area is the occupant load. Because buildings can have numerous functions, different occupant loads can be prescribed for different functions. The total capacity of all egress components must equal or exceed the occupant load. **[Ref. 1004.1]**
3. The code limits the distance from any point in the building to an exit. In some cases, the exit access travel distances can be increased by providing fire-resistance-rated construction or by installing an automatic sprinkler system. In most buildings, two or more means of egress are required, and exit doors and exit access doors must be spatially remote from one another. **[Ref. 1017.1]**
4. Exits, exit access and the exit discharge are constructed on horizontal planes. Any change of elevation is accomplished using ramps, stairs or steps. The code prescribes standard and consistent design criterion for these components to reduce and minimize tripping and fall hazards. **[Ref. 1011.1, 1012.1]**

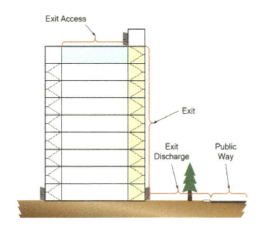

FIGURE 11-4 Exit access, exit and exit discharge

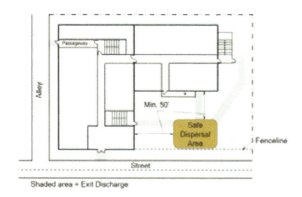

FIGURE 11-5 Safe dispersal area

5. The code requires illumination and identification of exit, exit access and exit discharge components. When two or more exits are required, emergency power for exit identification and illumination is required. [Ref. 1008.1, 1013.1, 1013.3]
6. To limit the potential for fire spread, the code prescribes more restrictive interior finish requirements to exit and exit access components when compared to the requirements for rooms or enclosed spaces. [Ref. 803.3]
7. Generally, all means of egress require accessibility for mobility-impaired individuals. Depending on the building's occupancy classification and level of fire protection, one or more areas of refuge may be required to shelter individuals who need rescue assistance. A means of communication is commonly required at the area of rescue assistance (Figure 11-6). [Ref. 1009.1]
8. A means of egress system that requires special knowledge or skills is not permitted. A code-compliant means of egress system simply requires persons to walk, operate simple door hardware and follow one of the designated paths to the level of discharge. [Ref. 1010.1, 1016.1, 1019.1]
9. Egress is complete when a person crosses the building property line and is in a public way. [Ref. 202]
10. Unless used for accessible means of egress, people movers such as elevators, escalators and moving sidewalks are prohibited from being considered as components of the means of egress system (Figure 11-7). [Ref. 1003.7]
11. The building code official is charged with approving the design and construction of new or renovated means of egress, including the occupant loads. The fire code official is responsible for ensuring the means of egress system is maintained.

FIGURE 11-6 This stairwell landing is designated as a location of rescue assistance for mobility-impaired persons.

FIGURE 11-7 An escalator and other types of people movers cannot be used as a component in a required means of egress system. The adjacent stairway provides the egress path.

OCCUPANT LOAD

Occupant load is the number of persons for which a means of egress is designed. With the exception of uses involving fixed seating commonly found in churches, stadiums, auditoriums and restaurants, occupant loads are based on a density value that considers the number of square feet required for each person given a building's function or use. The occupant load is calculated by dividing the floor area being considered by the occupant per unit area factor value specified in IFC Table 1004.5 (Table 11-1). Note that when applying this table, certain occupant load factors are based on the gross versus net floor area. Gross area is the area within the inside perimeter of exterior walls of the building and excludes vent shafts, columns and the thickness of interior walls. Net floor area is the actual occupied area and further excludes unoccupied accessory uses including corridors, stairways, toilets, mechanical rooms and closets. [Ref. 1004.1]

TABLE 11-1 Maximum floor area allowances per occupant (IFC Table 1004.5)

Function of space	Occupant load factor[a]
Accessory storage areas, mechanical equipment room	300 gross
Agricultural building	300 gross
Aircraft hangars	500 gross
Airport terminal	
Baggage claim	20 gross
Baggage handling	300 gross
Concourse	100 gross
Waiting areas	15 gross
Assembly	
Gaming floors (keno, slots, etc.)	11 gross
Exhibit gallery and museum	30 net
Assembly with fixed seats	See Section 1004.4
Assembly without fixed seats	
Concentrated (chairs only—not fixed)	7 net
Standing space	5 net
Unconcentrated (tables and chairs)	15 net
Bowling centers, allow 5 persons for each lane including 15 feet of runway, and for additional areas	7 net
Business areas	150 gross
Concentrated business use areas	See Section 1004.8
Courtrooms—other than fixed seating areas	40 net
Day care	35 net
Dormitories	50 gross
Educational	
Classroom area	20 net
Shops and other vocational room areas	50 net
Exercise rooms	50 gross
Group H-5 fabrication and manufacturing areas	200 gross
Industrial areas	100 gross
Institutional areas	
Inpatient treatment areas	240 gross
Outpatient areas	100 gross
Sleeping areas	120 gross
Kitchens, commercial	200 gross
Library	
Reading rooms	50 net
Stack area	100 gross
Locker rooms	50 gross
Mall buildings—covered and open	See Section 402.8.2 of the IBC
Mercantile	60 gross
Storage, stock, shipping areas	300 gross
Parking garages	200 gross
Residential	200 gross
Skating rinks, swimming pools	
Rink and pool	50 gross
Decks	15 gross
Stages and platforms	15 net
Warehouses	500 gross

For SI: 1 square foot = 0.0929 m^2, 1 foot = 304.8 mm.
a. Floor area in square feet per occupant.

In areas of buildings with fixed seating and aisles, the occupant load is calculated based on the number of seats. When the seats are not fixed, such as moveable tables and chairs, the occupant load is determined in accordance with the requirements in Section 1004.5. The calculated occupant load is added to the number of fixed seats that may be available. Calculation of the occupant load with fixed seating will also depend on the seat's design; if it is bench seating, 18 inches of seating space is calculated as one person and for dining areas, a seating space of 24 inches per person is applied. It should be noted that Table 1004.5 is based on the "function" of the space not the occupancy classification, as indicated in the heading for the left-hand column. This means that the occupant load factor of 20 net square feet for an educational classroom would apply in a Group E, Group B (college level with less than 50 occupants) or Group A-3 (college level with 50 or more occupants). [Ref. 1004.5, 1004.6]

> ### Code Essentials
> Occupant load is based on the function or purpose of the room or area. Depending on the function, it is calculated using either the net or gross floor area. In assembly occupancies, the maximum occupant load must be conspicuously posted. ●

EXERCISE

This example will demonstrate how the occupant load is determined in a room or building without fixed seating (refer to Figure 11-8).

Given: A single-story office building of 30,000 square feet

Determine: The occupant load for the building.

Solution: Occupant load factor = 150 ft²/person

(Table 1004.5)

30,000 ÷ 150 = 200

Occupant load = 200 persons

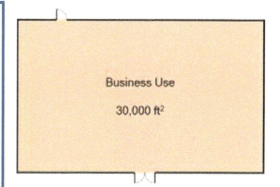

FIGURE 11-8 The occupant load factor is used to determine the occupant load of the space.

The occupant load must be posted in Group A occupancies (Figure 11-9). Posting the occupant load is essential for enforcement, especially when a fire department performs inspections of these occupancies during peak use. Many fire departments perform inspections of assembly occupancies at night or on weekends to ensure that the occupant load is not exceeded and that the egress system is functional. The occupant load sign must be conspicuously posted near the main exit or exit access door and be of a legible, permanent design. [Ref. 1004.9]

FIGURE 11-9 An occupant load sign

EGRESS WIDTH

Components of the means of egress must have an adequate width to safely accommodate the occupant load. Width is necessary to ensure the movement of the building occupants, and reductions in width require a reduction in the building occupant load. The width of means of egress components is specified in the code, such as the width of corridors or doors. In these cases, the width should be capable of safely accommodating the occupant load. To ensure that adequate capacity is built into the egress system, the code requires the calculation of egress width. The factors used in the calculation of exit width are dependent on the function of the means of egress component. For changes in grade or elevation that require the use of stairs, a factor of 0.3 inch/occupant is used (Figure 11-10). For all other egress components such as corridors, doors and ramps, a factor of 0.2 inch/occupant is used (Figure 11-11). In multistory buildings, the maximum capacity required from any story of the building must be maintained to the termination of the means of egress. [Ref. 1005.3, 1005.4]

FIGURE 11-10 The minimum width of an exit stair is 0.3 inch/occupant.

FIGURE 11-11 The minimum width of a corridor or ramp is 0.2 inch/occupant.

In all occupancies other than Groups H and I-2, the required width of means of egress components can be modified when an automatic sprinkler system and an emergency voice/alarm communication system are installed. The emergency voice/alarm communication system will provide early notification for occupants in the building thus allowing more time for a safe evacuation. The fire sprinkler system will act upon the fire to slow the fire progress, also allowing for a safer evacuation. When these two fire protection systems are installed, the stairway width is determined using a factor of 0.2 inch/occupant. All other egress components receive a 25 percent reduction and those widths are based on a factor of 0.15 inch/occupant. These reductions are allowed when the fire sprinkler system is designed according to NFPA 13 or 13R, as appropriate for the building. Keep in mind that the emergency voice/alarm communication system is not just a standard fire alarm system. This system consists of a fire alarm system, plus it must provide manual and automatic devices for originating and delivering voice instructions. See Chapter 10 for more information on emergency voice/alarm communication systems. [Ref. 1005.3]

Code Essentials

The exit width must at least equal the occupant load of the room, area or building. When required, at least two separate and remote paths of means of egress are provided. •

> **EXERCISE**
> This example will demonstrate how the required egress width can vary within an egress route. However, each component of the egress system is sized to adequately accommodate the occupant load served (refer to Figure 11-12).
> **Given:** A portion of the egress system serves 200 occupants in an unsprinklered Group B building.
> **Determine:** The minimum required widths of each egress component.
> **Solution:** Minimum Required Calculated Stair Width:
> Exit width per person for stairs = 0.3 inch/person (Sec. 1005.3.1) (200) × (0.3 inch/person) = 60 inches
> Minimum Required Calculated Width of All Other Components:
> Exit width per person = 0.2 inch/person (Sec. 1005.3.2) (200) × (0.2 inch/person) = 40 inches
> Table 1020.2 specifies a minimum corridor width of 44 inches when serving 50 or more occupants.

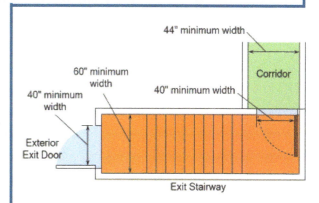

FIGURE 11-12 The width of each means of egress component is sized according to the occupant load it serves.

A properly designed means of egress system is commonly built with two separate and remote paths. The requirements for egress width follow this philosophy in that adequate width must be provided so that in the event one egress path is lost or compromised, the available width is reduced no more than 50 percent of the required capacity. **[Ref. 1005.5]**

If improperly installed, an egress door can create obstructions in the exit width (Figure 11-13). The IFC establishes requirements for doors that encroach into the egress path to ensure the needed width is not reduced. When fully open, doors cannot reduce the width of an egress component by more than 7 inches. In any other position, the door cannot reduce the required width by more than 50 percent. Surface-mounted latch release hardware, such as door knobs or release latches, are exempt from the 7-inch projection requirement if they face the corridor when the door is fully open. **[Ref. 1005.7.1]**

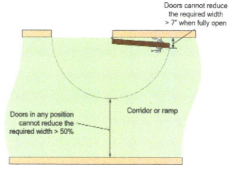

FIGURE 11-13 Door encroachment into an egress path in the corridor

EXIT ACCESS AND EXIT ACCESS TRAVEL DISTANCE

Another important component is the distance one must travel to safely egress a building. Maximum travel distances to reach an exit are established in the IFC. The exit access pathway must be clear and distinct and should avoid travel through intervening spaces that could confuse occupants attempting to egress a building. Travel distances and the geometry of exits and exit access are closely regulated in the code. The exit access requirement intends that exits should be direct from the room or area under consideration. However, this is not always possible, so the code grants exceptions for a limited number of circumstances where exits can pass through adjoining rooms or spaces rather than directly into corridors or exit enclosures.

Egress through an adjacent area or room is allowed provided the exit path is direct and obvious so the occupant is cognizant of the travel path. When egress is through an intervening space, it must be under the same control as the space where egress commenced, and the space under consideration must be accessory, meaning that its use is complementary to the use of the room or area where egress starts. The code is more concerned with the use of spaces rather than their size when evaluating this component of the means of egress. **[Ref. 1016.2]**

The exit access path cannot pass through a room or area that can be locked. In addition, certain building uses present a very high probability of being obstructed, including closets, storage rooms and kitchens. Accordingly, the code does not permit the use of these areas as exit access pathways. An exception is granted to Group M occupancies by allowing egress through stockrooms, because this occupancy consistently has a very low fire loss history and the occupants are cognizant and aware of their surroundings (Figure 11-14). However, there are specific limits, including that stored goods and commodities must have the same hazard classification as found in the main retail area, the egress path is demarcated by partial or full-height walls with a minimum width of 44 inches that leads to an exit, and no more than 50 percent of the exit access is through the stockroom area. In addition, doors into the stockroom cannot be locked. **[Ref. 1016.2]**

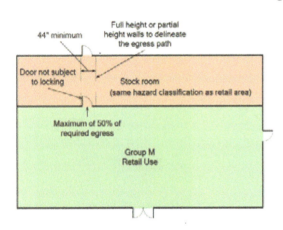

FIGURE 11-14 Exit access is allowed through the stockroom in a Group M occupancy provided it is constructed, available and maintained properly.

The distance occupants must travel along the exit access is based on the occupancy classification or a specific building's use, such as covered malls or atriums, and whether the building is equipped with an automatic sprinkler system. The requirements for travel distance are measured from any occupied point in the building to the nearest exit rather than to all required exits from a room, floor or building. Travel distance is measured from the most remote point in each room to the nearest exit. Exit access travel distances are shown in Table 11-2 and are measured to

- an exit passageway
- a horizontal exit
- the door to an interior exit stairway or ramp
- the door to an exterior exit stairway or ramp
- an exterior exit door located at the level of exit discharge **[Ref. 202]**

TABLE 11-2 Exit access travel distance (IFC Table 1017.2)

Occupancy	Without sprinkler system (feet)	With sprinkler system (feet)
A, E, F-1, M, R, S-1	200	250[b,e]
I-1	Not permitted	250[b]
B	200	300[c]
F-2, S-2, U	300	400[c]
H-1	Not permitted	75[d]
H-2	Not permitted	100[d]
H-3	Not permitted	150[d]
H-4	Not permitted	175[d]
H-5	Not permitted	200[c]
I-2, I-3	Not permitted	200[c]
I-4	150	200[c]

(Footnotes not shown. See IFC Table 1017.2 for all footnotes.)

Travel distance is measured along the natural and unobstructed path available to the occupants. The path will need to navigate furnishings such as workstations, chairs, machinery or fixtures such as shelves or storage racks. Some of these items may be moveable, which can complicate plan review or inspections. One conservative method that is reasonable in application is measuring using right angles, because it recognizes that obstructions such as machinery, racks and desks will need to be circumnavigated by occupants. This method of measurement accounts for any future obstructions that may be located in the exit access travel path and works well at plan review when the furnishings and equipment are not depicted. **[Ref. 1017.3]**

Except in occupancies with small areas, a properly designed egress system commonly has two or more exits or exit access doorways. The code requires two such doorways when:

- the occupant load exceeds the values in Table 1006.2.1
- the common path of egress travel exceeds the distance limitations in Table 1006.2.1
- the room or space contains a specific hazard—boilers, incinerators, furnaces, refrigeration machinery or refrigerated rooms.

Table 11-3 contains the maximum thresholds for occupant load and common path of egress travel which allow a single exit or exit access door. **[Ref. 1006.2]**

TABLE 11-3 Spaces with one exit or exit access doorway (IFC Table 1006.2.1)

Occupancy	Maximum occupant load of space	Maximum common path of egress travel distance (feet)		
		Without sprinkler system (feet)		With sprinkler system (feet)
		OL ≤ 30	OL > 30	
A c, E, M	49	75	75	75 a
B	49	100	75	100 a
F	49	75	75	100 a
H-1, H-2, H-3	3	NP	NP	25 b
H-4, H-5	10	NP	NP	75 b
I-1, I-2 d, I-4	10	NP	NP	75 a
I-3	10	NP	NP	100 a
R-1	10	NP	NP	75 a
R-2	20	NP	NP	125 a
R-3 c	20	NP	NP	125 a, g
R-4 c	20	NP	NP	125 a, g
S f	29	100	75	100 a, g
U	49	100	75	75 a

For SI: 1 foot = 304.8 mm. NP = Not Permitted.
a. Buildings equipped throughout with an automatic sprinkler system in accordance with Section 903.3.1.1 or 903.3.1.2. See Section 903 for occupancies where automatic sprinkler systems are permitted in accordance with Section 903.3.1.2.
b. Group H occupancies equipped throughout with an automatic sprinkler system in accordance with Section 903.2.5.
c. For a room or space used for assembly purposes having fixed seating, see Section 1029.8.
d. For the travel distance limitations in Group I-2, see Section 407.4 of the *International Building Code*.
e. The common path of egress travel distance shall apply only in a Group R-3 occupancy located in a mixed occupancy building or within a Group R-3 or R-4 congregate living facility.
f. The length of common path of egress travel distance in a Group S-2 open parking garage shall be not more than 100 feet.
g. For the travel distance limitations in Group R-3 and R-4 equipped throughout with an automatic sprinkler system in accordance with Section 903.3.1.3, see Section 1006.2.2.6.

Beyond the allowance for one exit or exit access doorway, the IFC not only requires an increase in exit width as the occupant load increases, but also prescribes a greater number of exit or exit access doorways. The code allows two exits to serve up to 500 occupants. From 501 to 1,000 occupants, three separate exits are required. When the occupant load exceeds 1,000, a minimum of four exits are required. The example below combines the requirements for minimum exit width with the number of exits. **[Ref. Table 1006.3.2]**

EXERCISE

Given: Fully sprinklered Group M occupancy of 62,000 square feet. Refer to Figure 11-15.

Determine: The minimum number of exits and the minimum exit width

Solution:

Occupant load
Table 1004.1.2 requires 60 ft²/person
62,000 ÷ 60 = 1,033 occupants

Number of exits
Table 1006.3.2 requires 4 exits
4 exits provided—**OK**

Exit width
1,033 × 0.2 = 207" required
An exit door must provide 32" clear width
7 exit doors provided
7 × 32 = 224" of exit width—**OK**

FIGURE 11-15 Example for determining required exit width and number of exits

The IFC prescribes specific requirements relative to the location and arrangement of the required exit or exit access doorways in relation to one another. These requirements ensure a higher degree of reliability in that if one means of egress is obstructed, the others will remain available and will be usable by the occupants. This approach assumes that since the remaining means of egress are still available, there will be sufficient time for the building occupants to use them to evacuate the building or space. If two exits or exit access doorways are required, they must be arranged and separated to provide access from the entire space, room or building served. The minimum distance between the two means of egress, measured in a straight line, shall not be less than one-half of that maximum overall diagonal dimension of the room or space (Figure 11-16). Measurements are taken in a straight line as if looking at a set of plans. **[Ref. 1007.1.1]**

In buildings protected throughout by an automatic sprinkler system complying with NFPA 13 or NFPA 13R, the separation of the exit or exit access doorways can be reduced (Figure 11-17). Because the building is protected by an automatic sprinkler system, the IBC and IFC allow the distance between exits and exit access doorways to be reduced to one-third instead of the conventional one-half value prescribed for all buildings not protected by sprinkler systems. **[Ref. 1007.1.1, Exception 2]**

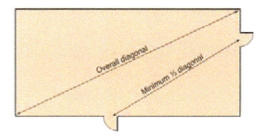

FIGURE 11-16 Where two exits or exit access doorways are required, they must be separated by one-half of the longest diagonal dimension of the space served. Measurements are made in a straight line.

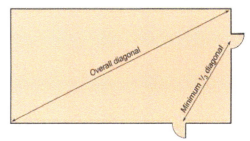

FIGURE 11-17 The separation between exit and exit access doorways can be reduced to one-third of the overall diagonal in buildings sprinklered in accordance with NFPA 13 or NFPA 13R.

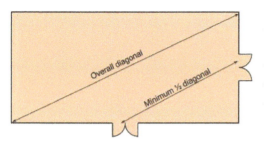

FIGURE 11-18 The separation between exit doorways is measured at any point along the doorway width.

The code specifies how the measurement of the separation distance between exits or exit access doors is to be made. Considering that a pair of exit doors is typically 6 feet wide, the point of measurement along the doors can affect the location of the doors in the room or building. The IFC specifies that separations shall be measured as follows:

1. For exit or exit access doorways, measure to any point along the width of the doorway
2. For exit access stairways, measure to the closest riser
3. For exit access ramps, measure to the start of the ramp run (Figure 11-18) [Ref. 1007.1.1.1]

EXIT SIGNS AND MEANS OF EGRESS ILLUMINATION

An unwanted fire, natural or technological disaster can impact the life safety of building occupants. One concern is the ability to visually locate the means of egress and clearly identify openings that allow occupants to quickly and easily pass through the level of exit discharge. The code requires illumination of the means of egress when the building is occupied and identification of exit and exit access doorways using signs constructed to specific standards that can be easily recognized and comprehended by the general population. With the exception of Group U occupancies, aisle accessways in Group A occupancies, dwelling and sleeping units in Group I, R-1, R-2 and R-3 occupancies, means of egress illumination is required for all occupancies. Illumination must be provided throughout the exit access, exit and at the point of exit discharge. [Ref. 1008.2]

Exit illumination must emit enough visible optic energy to equal 1 footcandle at the walking surface. One footcandle of luminance equals 10.76 lux, which is the SI-derived unit of measurement. Lux is commonly expressed as lumens, which is the amount of light the source produces and approximately equals 0.1 footcandle. One lux equals one lumen per square meter of area and 1 footcandle equals 1 lumen illuminating one square foot. Eleven lux (or lumens) equal 1 footcandle. A 40-watt light bulb emits about 455 lumens or about 41.4 footcandles. [Ref. 1008.2.1]

While the probability of utility electric power failure is low, its failure is a credible event, and the IFC specifies requirements to ensure an alternative electrical power source is available for means of egress and exit sign illumination (Figure 11-19).

Emergency power, which must operate within 10 seconds of power loss, is required for building egress components that require two or more exit or exit access doorways. Emergency power for egress illumination is required in aisles, corridors, exit enclosures and exit passageways, as well as exterior landings at the exit discharge. Specific rooms or areas require emergency power regardless

FIGURE 11-19 Emergency lighting for the means of egress is activated when normal power is lost.

of the number of exits required. These rooms are fire command centers, fire pump rooms, generator rooms, electrical rooms and public restrooms larger than 300 square feet. [Ref. 1008.3]

The source of emergency power must be capable of operating for a minimum of 90 minutes. The source of power can be storage batteries, unit equipment that is equipped with its own standby power source or an engine-driven generator. Regardless of the power source, it must be maintained in accordance with IFC Section 1203. (See Chapter 6 of this text for a review of the IFC, IBC and NFPA requirements for standby and emergency power systems.) [Ref. 1008.3.4]

To clearly demarcate exit and exit access doorways, the IFC requires the installation of approved exit signs. Signs are required to indicate the direction for egress travel when the exit or exit access is not clearly or immediately visible to building occupants. Signs are not required in rooms or areas with only one exit or exit access door and, when approved by the code official, in buildings where the main exterior exit doors are obvious and identifiable. Signs must be placed in accordance with their listed viewing distance but not more than 100 feet from where an exit or exit access doorway enters a corridor (Figures 11-20 and 11-21). Exit signs are not required in Group U occupancies and in sleeping or dwelling units of Group R-1, R-2, R-3 and I-3 occupancies. [Ref. 1013.1]

Exit signs can be either internally or externally illuminated. Internally illuminated signs can use electric lamps or chemicals that are self-illuminating or photoluminescent. All exit signs must be listed and labeled as complying with UL 924, Standard for Emergency Lighting and Power Equipment (Figure 11-22). Internally illuminated signs must remain illuminated at all times. In the event of a power loss, UL-listed exit signs constructed as unit equipment are required to operate for at least 90 minutes at no less than 60 percent of their rated voltage. [Ref. 1013.3]

Externally illuminated exit signs also are allowed. When externally illuminated signs are used, they must meet the dimensional sizing requirements and be of a contrasting color. At the face of the sign, there must be a minimum luminance level of not less than 5 foot-candles. The source of lighting must be connected to an approved emergency power source that can provide electricity for at least 90 minutes. [Ref. 1013.6]

> **You Should Know**
>
> Means of egress is a critical component providing life safety in buildings. The IFC addresses all portions of the design, operation, testing and maintenance of doorways, stairways and ramps that provide a portion of the means of egress.

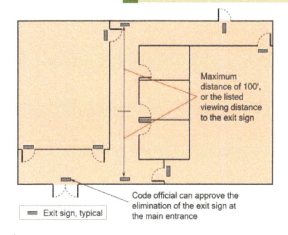

FIGURE 11-20 Exit signs are required in each room or space required to have two or more exits. The viewing distance is measured from where the occupant would enter the corridor.

FIGURE 11-21 This listed exit sign includes a marking to indicate that its maximum viewing distance is 50 feet.

FIGURE 11-22 UL 924-listed internally illuminated exit sign

FIGURE 11-23 Obstructions on the exterior side of the exit door, such as shopping carts blocking the door from opening, are a serious fire code violation.

MEANS OF EGRESS MAINTENANCE

IFC Section 1031 establishes requirements for the maintenance of the means of egress system. These provisions address obstructing, blocking or disabling means of egress components. The requirements also are intended to prevent conditions that can cause confusion or obscure the means of egress.

When a building is occupied, regardless of occupant load, the exit components are required to be maintained free of any obstructions or impediments that can prevent the complete and instantaneous use of the means of egress. All components of the means of egress must be available in the event of an emergency that requires occupants to leave a building. The IFC prohibits any obstructions to exits, exit access or the exit discharge. Such obstructions must be immediately removed (Figures 11-23, 11-24 and 11-25). **[Ref. 1031.2, 1031.3]**

FIGURE 11-24 Objects on the interior side of the exit door, such as the corridor being used for storage, obstruct access to the exit and add fire load into the exit path.

FIGURE 11-25 Security measures on the interior side of the exit door, such as a chain on the panic hardware, is a serious fire code violation.

The IFC requires exit signs to be installed in accordance with Chapter 10 requirements. Exit doors, exit access doors and exit signs must not be obstructed or blocked by drapes, decorations or partitions, nor should there be other signs that can distract or cause confusion (Figure 11-26). **[Ref. 1030.4, 1031.6]**

FIGURE 11-26 Objects, signs, or too many exit signs that confuse the egress path are prohibited. Should the occupant go straight forward, or turn left as indicated in the closest exit sign?

Means of Egress Maintenance 161

EXERCISE

Given: Loretta's Honky Tonk is a Group A-2 night club that serves alcoholic beverages. The business is moving into an existing wood frame building previously used as a convenience store. The building is protected throughout by an automatic sprinkler system in accordance with IFC Section 903.3.1.1. The building will be renovated by the addition of fixed seating and a dance floor. (See Figure 11-27)

An application was submitted to the Fire Department for a Group A occupancy operating permit.

Determine:

1. The occupant load based on the amount of open space in the occupancy
2. The fixed seating occupant load
3. The required exit width based on the occupant load.

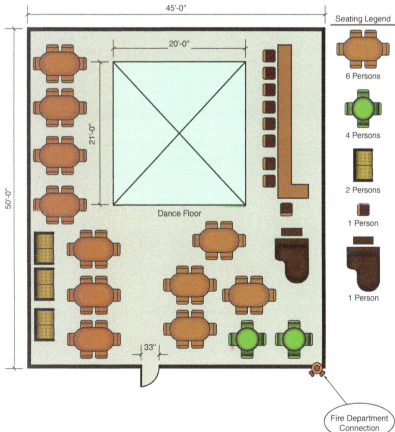

Loretta's Honky Tonk

Calculation Notes:
- Calculate the occupant load based on the number of fixed seats and the area of the dance floor.
- The occupant load factor for the dance floor is 7 sq. ft. / person.
- The egress width factor for the exit door is 0.2 inches / person
- The serving area behind the bar is 51 square feet and the exit aisle from the dance floor is 3 feet wide by 28 feet length.

FIGURE 11-27 Means of egress exercise

Solution: Occupant load calculation

Fixed seats	Number	Occupants
Tables for 6 persons	11	66
Tables for 4 persons	2	8
Tables for 2 persons	3	6
Individual chairs/stools	10	10
TOTAL FIXED SEATS		90

Dance floor occupant load: (420 ft.^{2}) ÷ (7 ft.^{2}/person)* = 60 persons

Bar service area occupant load: (51 ft.^{2}) ÷ (200 ft.^{2}/person)** = 1 person

Exit aisle occupant load: (168 ft.^{2}) ÷ (5 ft.^{2}/person)*** = 34 persons

Total occupant load: 90 + 60 + 1 + 34 = 185 persons

*Based on IFC Table 1004.5 value for assembly—concentrated use

**Based on IFC Table 1004.5 value for commercial kitchens

***Based on IFC Table 1004.5 value for assembly—standing area

Exit width calculation

185 occupants × 0.2 inch/person = required clear width of 37 inches

Findings

1. The exit width is inadequate for the proposed use. **[Ref. 1005.3.2]**
2. Based on the occupant load, a second exit or exit access doorway is required. **[Ref. Table 1006.2.1]**
3. Panic hardware is required for the exit doors. **[Ref. 1010.1.10]**
4. Main door could have a key-operated lock rather than panic hardware. **[Ref. 1010.1.9.4]**
5. Exit signs are required at each exit door. **[Ref. 1013.1]**

PART V: Special Processes and Building Uses

Chapter 12: Motor Fuel-Dispensing Facilities and Repair Garages

Chapter 13: Flammable Finishes

Chapter 14: High-Piled Combustible Storage

Chapter 15: Other Special Uses and Processes

CHAPTER 12
Motor Fuel-Dispensing Facilities and Repair Garages

Motor fuel-dispensing facilities and repair garages are closely regulated because they allow the general population to handle flammable and combustible liquids and flammable gases that are more hazardous than the public may realize. In all states but New Jersey and portions of Oregon, the public can perform their own fuel dispensing at public-accessible fueling facilities. The dispenser is often the point of sale for the entire dispensing operation. To minimize the risk of fire or unauthorized discharge of hazardous materials, Chapter 23 of the *International Fire Code* (IFC) sets forth requirements based on the hazards of the various fuels available in the marketplace. Regardless of the fuel, dispensing equipment and components accessible and used by the motoring public must be listed by a nationally recognized testing laboratory.

APPLICABLE REQUIREMENTS BY FUEL

IFC Chapter 23 provisions are based on the type and hazards of the fuel. They are all flammable or combustible, can be a liquid or gas and the gases can be compressed, liquefied or cryogenic (refrigerated gases). The IFC prescribes specific requirements because of the characteristics of the particular fuel and its hazards. Chapter 23 addresses storage and dispensing of fuels like unleaded gasoline, alcohol-gasoline mixtures and petroleum distillates, such as No. 2 diesel fuel, kerosene and biodiesel. It also addresses liquefied petroleum gases (LP-gas), natural gas (methane) and hydrogen stored as a compressed gas or cryogenic fluid. The provisions for these fuels are applied in conjunction with the applicable hazardous material requirements in IFC Chapters 53, 55, 57, 58 and 61 and the applicable NFPA standards. Regardless of the fuel and its physical state, all motor fuel-dispensing stations must comply with the general provisions in Section 2301 and the adopted NFPA standards indicated in Table 12-1.

> **Code Essentials**
>
> Motor fuel dispensing requirements apply to fuel storage and dispensing operations in automobiles, water craft, aircraft, etc; essentially any motor vehicle with a fixed fuel tank.

TABLE 12-1 Applicable IFC requirements and NFPA standards by fuel

Stored and dispensed fuel	Applicable IFC requirements	Applicable NFPA standards
Gasoline, diesel fuel, gasoline/ethanol mixtures	Section 2306, Chapter 57	NFPA 30, NFPA 30A
Hydrogen – Compressed	Section 2309, Section 5303	NFPA 55
Hydrogen – Liquefied	Section 2309, Section 5806	NFPA 55
Liquefied Petroleum Gas	Section 2307, Chapter 53; Chapter 61	NFPA 58
Natural Gas – Compressed	Section 2308, Chapter 53	NFPA 52
Natural Gas – Liquefied	Chapter 58	NFPA 55, NFPA 57, NFPA 59A

DISPENSING OPERATIONS AND DEVICES—ALL FUELS

Dispensing is "the pouring or transferring of any material from a container, tank, or similar vessel whereby vapors, dusts, fumes, mists or gases are liberated to the atmosphere." Dispensing operations under the IFC may be at- tended or unattended. Attended dispensing occurs where an individual at the motor fuel-dispensing station supervises or performs the transfer of fuel. Note that in "attended" dispensing operations, the attendant does not necessarily need to fuel the vehicle. The attendant must be able to supervise the fueling operations that occur. Self-service operations, where the fueling is performed by the customer, can occur at either an attended dispensing facility or unattended dispensing facility. A self-service operation with an attendant present who can supervise the dispensing operation, even though the attendant does not perform the fueling, is considered an attended dispensing operation. [Ref. 202]

FIGURE 12-1 This facility is considered an attended motor fuel-dispensing operation even though the attendant may not be fueling the vehicles.

FIGURE 12-2 The attendant must be able to view the dispensing operation.

FIGURE 12-3 There are no employees or assistance at unattended fuel-dispensing operations.

At attended dispensing operations, at least one responsible person is on site who supervises, controls and observes the fuel dispensing operations. This individual is responsible for ensuring containers filled with fuel are acceptable portable containers. Acceptable portable containers are constructed of listed or approved material with a maximum capacity of 6 gallons, and have tight closures with screwed or spring-loaded covers so designed that the contents can be dispensed without spilling. The attendant is responsible for controlling spills and is trained in the use of portable fire extinguishers. The attendant must be able to communicate with persons in the dispensing area and have access to the emergency controls to halt the dispensing operation if an incident occurs (Figure 12-1). **[Ref. 2304.2, 2304.4.1]**

At attended fuel-dispensing operations, the fueling can be performed by the attendant or the customer. A typical minimart facility consists of fueling islands and a small market. The fueling attendant is inside the market at the cash register. Some of the attendant's duties are to observe the fueling operation and prevent the filling of non-approved containers. In these facilities, these functions are facilitated by the IFC requirements that the attendant must be able to observe all of the fueling operations and be able to communicate with the customers at the fueling island. If the cash register is behind windows that face the fueling islands, the attendant will be able to view the fueling operations. The method of communication must be approved by the fire code official. These design criteria can be met and tested at the time of final approval of the permit. One of the popular advertising practices in these minimarts is to place posters and ads on the windows (Figure 12-2). This becomes an issue when those posters block or impair the attendant's complete view of the fueling islands. The visible perspective of the attendant is an item that must be inspected during the lifetime of the business operation. **[Ref. 2304.2.4, 2304.2.5]**

Unattended fuel dispensing is self-service dispensing and commonly the dispenser serves as the point of sale (Figure 12-3). The IFC requires the owner to perform a daily reconciliation of fuel sales as well as an inspection of dispensing equipment. The daily reconciliation verifies that the fuel storage tanks are not leaking or otherwise losing product. Each dispenser requires operating instructions and identification of the location of

Dispensing Operations and Devices—All Fuels

emergency controls statements for motor fuel-dispensing that can be used to stop dispensing operations, as well as statements that address dispensing of fuels into unauthorized containers. The signs inform individuals dispensing fuel of the actions they need to take in the event of a fire, spill or release (Figure 12-4). An on-site means of communicating with the fire department is required. **[Ref. 2304.3, 2305.6]**

Dispensing devices must be properly located in relation to property lines, buildings and their openings, and fixed sources of ignition. Dispensers are to be located so the vehicle receiving fuel and the dispensing nozzle are located on the same property. Regardless of the fuel type, all dispensers require a clearly identified emergency disconnect switch that stops dispensing upon activation (Figure 12-5). Switches can be located indoors or outdoors. At outdoor dispensing operations, the switch must be a minimum of 20 feet but not more than 100 feet from a dispenser. The 20-foot minimum separation protects individuals who are self-dispensing by removing the person from potential harm if a fuel spill is ignited. When dispensers are located indoors, the location of the emergency disconnect switch must meet the approval of the fire code official. **[Ref. 2303.1, 2303.2]**

The IFC also establishes operational requirements for tank filling, maintenance of dispensers and the control of spills and ignition sources (Figure 12-6). Before a storage tank is filled, the tank vehicle driver is required to gauge (measure) the tank and determine the available capacity before transferring product. The liquid transfer and vapor-recovery connections on storage tanks with a volume of more than 1,000 gallons must be liquid and vapor tight. When liquid is pumped into above-ground storage tanks, the tank vehicle must be located at least 15 feet from a tank receiving Class II and IIIA combustible liquids and 25 feet from a tank receiving a Class I flammable liquid. **[Ref. 2305.1]**

FIGURE 12-4 Operating instructions and warning labels are required on the dispenser.

FIGURE 12-5 An emergency disconnect switch must be readily accessible.

FIGURE 12-6 Connections from a tank vehicle to a storage tank must be liquid and vapor tight.

FIGURE 12-7 Devices such as these vapor and liquid leak detectors in a below-grade vaulted storage tank must be tested annually.

FIGURE 12-8 Portable fire extinguishers serving dispensers, pumps and tank-fill connections require a minimum 20-B Class B fire hazard rating. *(Courtesy of TYCO/Ansul, Marinette, WI)*

Dispensing equipment must be properly maintained so it does not become a source of ignition or a leak. Any equipment that is leaking should be removed from service as it is a well-known fact that leaks never become smaller. Emergency shutoff valves and liquid leak detectors require an annual functional test (Figure 12-7). When repairs are performed on dispensing devices, the IFC requires the electrical power to the dispenser and its source pump be disconnected, the dispenser emergency shutoff valve is closed and a minimum 12-foot exclusion zone be established. Only persons knowledgeable in performing the repairs are allowed within the work zone. **[Ref. 2305.2.3, 2305.2.4]**

Portable fire extinguishers are required near dispensers, pumps and storage tank fill connections. Minimum 2-A:20-B:C portable fire extinguishers are required within 75 feet of the indicated components. Portable fire extinguishers must be installed in accordance with IFC Section 906 and NFPA 10 (Figure 12-8). **[Ref. 2305.5]**

FLAMMABLE AND COMBUSTIBLE LIQUID FUEL DISPENSING

Unleaded gasoline, biodiesel, No. 2 diesel fuel and alcohol blended gasoline are commonly dispensed liquid fuels. Section 2306 addresses the storage, transfer, dispensing and vapor-recovery of these liquids. Because stationary storage tanks are used, their installation must comply with the requirements in IFC Chapter 57; however, the type of tank selected and its siting requirements must comply with the requirements in IFC Chapters 23 and 57. (A complete review of storage tank design and installation requirements is found in Chapter 18 of this text.) **[Ref. 2306.2]**

A common method of storing fuels is underground storage tanks (USTs). USTs are desired by many petroleum marketers because they eliminate the fire hazard of above-ground storage and do not consume any land area since the tank is buried. Because the tank is underground, leaks cannot be readily observed. To monitor for potential leaks, a means of leak detection is required for USTs storing Class I, II and III flammable and combustible liquids. However, the IFC imposes an additional requirement for daily product reconciliation between product sold, product received and the available inventory in the UST. A daily reconciliation is required for each storage tank and any consistent or accidental loss of petroleum product must be reported to the fire code official. In many cases, this reconciliation is performed automatically by the tank's inventory control equipment (Figure 12-9). **[Ref. 2306.2.1.1]**

FIGURE 12-9 Electronic UST inventory and leak detection device

Because of the general public's proximity to a fuel dispenser during dispensing and the potential for vehicular impact, the IFC requirements for above-ground storage tanks (AST) supplying fuel dispensers are more restrictive than the requirements for USTs. These increased requirements are justified because of the increased potential for fire or explosion. Class I flammable liquids, such as unleaded gasoline or alcohol-based fuels formulated with 85 percent ethanol/15 percent gasoline (E85) must be stored in a listed protected above-ground storage tank (PAST) (Figure 12-10). This particular type of AST is listed to the requirements of UL 2085, Standard for Safety Protected Aboveground Tanks for Flammable and Combustible Liquids, and is designed to resist damage from a hydrocarbon pool fire. The design and test criteria for PASTs were developed with the intent of creating an above-ground tank that provided an equivalent level of fire safety as an underground tank. With this in mind, PASTs are constructed with integral secondary containment, bullet-resistance and vehicle impact protection. Class II and IIIA combustible liquids such as No. 2 diesel fuel also must be stored in a PAST; however, the fire code official can permit the use of other ASTs for these liquids, since they have a higher flash point temperature, making them more difficult to ignite. The maximum volume for a single AST at a publicly-accessible fueling facility is 12,000 gallons, and the aggregate volume at the site is 48,000 gallons. Biodiesel is typically a Class IIIB combustible liquid. Because of its higher flash point, any tank listed for flammable or combustible liquid can be used. All tanks must be located in accordance with the requirements in Table 2306.2.3, which are more restrictive than the tank siting requirements in Chapter 57. [Ref. 2306.2.3]

FIGURE 12-10 This protected above-ground storage tank contains gasoline, which is a Class IB flammable liquid.

Another method that can increase the stored volume is a vaulted tank. Vaulted tanks can be installed above grade or below grade (Figure 12-11). A vaulted tank utilizes a conventional, non-insulated AST. The storage tank is placed inside a concrete vault that has a concrete cover. Such a design allows for an AST to be installed below grade, which can benefit the owner because these tanks are exempt from required payments into the jurisdiction's leaking storage tank fund. The liquid-tight vault serves as secondary containment

Code Essentials

Storage tanks for fuel dispensing of Class I flammable liquids must be either
- Underground storage tanks
- UL 2085 listed protected above-ground storage tanks
- Tanks located in an above-grade or below-grade vault.

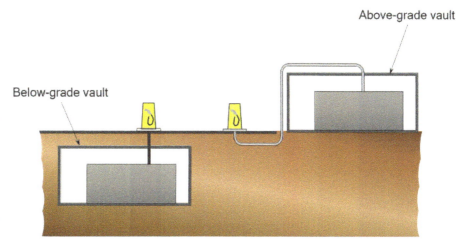

FIGURE 12-11 Tanks in a vault can be installed below grade or above grade.

FIGURE 12-12 Interior of an above-ground tank in a below-grade vault

FIGURE 12-13 Overfill prevention device *(Courtesy of OPW Fueling Components, Dublin, OH)*

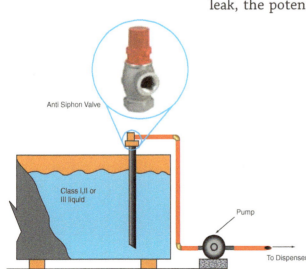

FIGURE 12-14 Anti-siphon valve

if a leak occurs in the primary tank. Vaulted tanks are listed as meeting UL 2245, Below-Grade Vaults for Flammable Liquid Storage Tanks, and require a mechanical ventilation system, overfill prevention device and a means of vapor and liquid leak detection. Vaulted ASTs are permitted to store up to 15,000 gallons of Class I, II and IIIA liquids (Figure 12-12). For fuels classified as Class IIIB, a tank designed to either UL 142, Steel Aboveground Tanks for Flammable and Combustible Liquids or a PAST is allowed. **[Ref. 2306.2.4]**

To reduce the potential for leaks, the IFC has requirements for the design of the piping and valve systems to ensure that if a leak occurs, the amount released is minimized. These provisions are used in conjunction with the requirements of IFC Chapter 57.

All tank openings are required to be through the top of the storage tank. By requiring openings at the top of the tank, all liquids must be pumped to the dispenser, which is safer than using a gravity delivery system. Such a system ensures that liquid is below any openings, limiting the potential for leaks. To prevent the overfilling of an AST, the IFC requires the installation of an overfill prevention device. These devices are installed in the tank fill pipe connection and are designed to limit the flow of liquid when the tank is filled to 90 percent of its volume. When the tank is filled to 95 percent of its volume, the overfill prevention device closes and prevents the delivery of any additional petroleum product (Figure 12-13). **[Ref. 2306.6.2.3]**

Because the IFC requires all openings for tanks to be through the top of the tank, transfer pumps are required. Pumps can be installed at the top of the tank or at the dispenser. When a pump is installed at the dispenser, the potential for product siphoning can occur from above-ground tanks. If the pump is operated and the piping has a leak, the potential exists for the negative pressure to siphon liquid from the storage tank, which can increase the size of the release. When such designs are used, the IFC requires the installation of an anti-siphon valve. This can be a solenoid valve that opens when the dispenser is activated or the installation of a valve that is designed to prevent siphoning of product in the dispenser supply pipe (Figure 12-14). **[Ref. 2306.6.2.4]**

The fuel dispenser is the point of use by the consumer. Because the predominance of fuel dispensing is performed in a self-service setting, the IFC specifies requirements to minimize the potential of a spill and fire. These requirements address the method of fuel delivery, mounting and construction of dispensers and the hose and nozzles.

All components of a flammable or combustible liquid-dispensing system are required to be listed by a nationally recognized testing laboratory. The IFC requires listed electrical equipment, dispensers, hoses, nozzles and submersible or subsurface pumps used for the movement and dispensing of liquid. Listing of equipment ensures that it has been evaluated for the fuel being dispensed and is constructed and assembled using materials compatible with the fuel (Figure 12-15). [Ref. 2306.7.1]

FIGURE 12-15 Variable frequency drive pump controllers. These devices control the flow rate of pumps supplying dispensers. Because they are part of the dispensing system, these controllers must be listed in order to comply with the IFC.

A major concern with any dispenser is vehicular impact. Dispensers are equipped with valves and fittings that prevent the release of flammable and combustible liquids if it is impacted by a vehicle. Protection from vehicular impact can be accomplished by the installation of guard posts or the installation of the dispensers on a 6-inch-high concrete island or any other barrier approved by the fire code official (Figure 12-16). The IFC requires the installation of a dispenser emergency shutoff valve at the base of the dispenser in each product line that supplies a dispenser. The valve is equipped with a low melt point fusible link so in the event of a fire at the dispenser, the link will fail and close the valve. Dispenser emergency shutoff valves are required to be functionally tested annually (Figure 12-17). [Ref. 312, 2305.2.4, 2306.4, 2306.7.3, 2306.7.4]

A dispenser emergency shutoff valve has a shear section that will separate from the main body of the valve attached to the dispenser base. Separation of the shear section mechanically stops the flow of liquid. Positioning of the shear section is important, because it must be parallel to the dispenser foundation and properly located so the shear section can effectively operate when impacted. The IFC requires it be installed within ½-inch of the top of the island. Equally important is that the valve be installed in accordance with its listing (Figure 12-18). [Ref. 2306.7.4]

FIGURE 12-16 Dispensers protected from vehicular impact

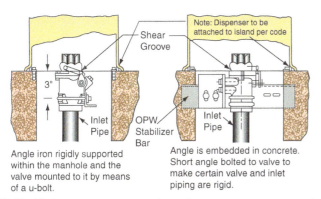

FIGURE 12-18 Proper installation of the dispenser emergency valve should include verification of the location of shear section in relation to the foundation. (Courtesy of OPW Fueling Components, Dublin, OH)

FIGURE 12-17 An automatic emergency shutoff valve is installed at the base of the dispenser. (Courtesy of OPW Fueling Components, Dublin, OH)

FIGURE 12-19 An automatic emergency valve for vapor-recovery lines in the dispenser *(Courtesy of OPW Fueling Components, Dublin, OH)*

FIGURE 12-20 An emergency breakaway device is required in the dispensing hose for Class I and Class II liquids.

The U.S. Environmental Protection Agency can mandate the installation of vapor-recovery and vapor-processing systems to capture vapor that can be released during the dispensing of unleaded gasoline and ethanol/gasoline mixtures. Vapor-recovery piping is installed as a part of the dispenser, so an emergency shutoff valve is required (Figure 12-19). **[Ref. 2306.7.4]**

Another safety component installed on a dispenser is an emergency breakaway device that is designed to safely separate the dispensing hose if a vehicle drives away when the fuel-delivery nozzle is attached to the vehicle (Figure 12-20). An emergency breakaway device is required on all dispensers conveying Class I and II liquids. When operated, the emergency breakaway device will contain liquid on both sides of the hose that is separated. **[Ref. 2306.7.5.1]**

The point of use for motor-fuel dispensers is the fuel-delivery nozzle. Fuel-delivery nozzles dispensing Class I, II or III liquids are required to be automatic closing. The fire code official can dictate when a latch-open device is or is not allowed. Some code officials do not allow latch-open devices because it requires the person dispensing the fuel to be in constant attendance of the delivery nozzle. Conversely, other fire code officials require latch-open devices because they prevent the person dispensing fuel from inserting objects between the nozzle body and valve, such as vehicle fuel caps or wallets. The IFC specifies that if a latch-open device is used it must be an integral component of the listed nozzle (Figure 12-21). **[Ref. 2306.7.6.1]**

FIGURE 12-21 This listed fuel-delivery nozzle is equipped with a latch-open device.

Fuel-delivery nozzles are designed so they will only allow dispensing when the dispenser hose is pressurized with liquid. Loss of pressure, such as turning off a pump, will cause the nozzle valve to close. The nozzle must be manually operated to ensure it is closed before dispensing can be resumed. Its design must have a feature or component that retains the nozzle in the vehicle fill pipe while dispensing is underway. **[Ref. 2306.7.6.1]**

Alcohol-blended fuel is defined in Section 202; the definition specifies common fuel formulations. Ethanol is a water-miscible solvent classified as a Class IB flammable liquid because of its closed cup flash point temperature of 55°F and a boiling point temperature of 173°F. Ethanol (designated as "E") is blended with gasoline to make up two different formulations of alcohol-blended fuels: E10, which is a blend of 10 percent ethanol with 90 percent unleaded gasoline, and E85, which is a blend of 85 percent ethanol with 15 percent gasoline. E85 alcohol-blended fuel has a closed cup flash point temperature ranging from -20 to -4°F, which is higher than unleaded gasoline, which has a -45°F flash point temperature. The flammable range of E85 alcohol-blended fuel is much greater than that of unleaded gasoline as indicated in Table 12-2.

TABLE 12-2 Flammability ranges of unleaded gasoline and alcohol-blended fuel

Fuel	Lower flammable limit (percent volume in air)	Upper flammable limit (percent volume in air)	Flammable range (percent volume in air)
Unleaded gasoline	1.4	7.6	6.2
E85 alcohol-blended fuel	1.4	19.0	17.6

The IFC requires listed components for the dispensing of alcohol-blended fuels, such as nozzles, hoses, pumps, breakaway connections or any other component that will be wetted by the fuel. Compared with gasoline, ethanol and ethanol-blended fuels have increased electrical conductivity, which can affect material compatibility with a potential effect of increased corrosion. In this context, compatibility is defined as the ability of two or more substances to maintain their respective physical and chemical properties on contact with one another for the design life of the storage and dispensing system under conditions likely to be encountered. Alcohol-blended fuels will aggressively attack and cause premature failure of components constructed of metals, alloys and plastics that are not chemically compatible with ethyl alcohol. Components constructed of soft metals, including zinc, brass, aluminum or lead, are chemically incompatible with E85 alcohol-blended fuel. Unplated carbon steel, stainless steel and bronze are resistant to ethanol. Nonmetallic materials that are compatible with alcohol-blended fuels include neoprene rubber, polypropylene, nitrile plastic and polytetrafluoroethylene (registered under the trademark Teflon).
[Ref. 2306.8.1]

If a motor vehicle fuel-dispensing facility switches fuels from unleaded gasoline to an alcohol-formulated fuel using the same storage tanks and dispensing equipment, the conversion is subject to a review and approval by the fire code official. Ethanol is an extremely water-miscible liquid, and if equipment is not properly cleaned or prepared for the storage of alcohol-blended fuels, the ethanol will absorb any water and contaminate the fuel. If the facility stores petroleum fuels in underground storage tanks constructed of fiberglass-reinforced plastic, the tank may or not be compatible with E85

FIGURE 12-22 An E85 alcohol-unleaded gasoline dispenser

fuel. Since the mid-1980s, all fiberglass-reinforced plastic underground storage tanks constructed with integral secondary containment (such as a double-wall) are compatible with 100-percent ethyl alcohol. In each case it is important that the original tank installation records be reviewed and any questions concerning the compatibility of the tank with the fuel be referred to the manufacturer. Automatic tank liquid level gauges using capacitance probes will not work with alcohol-blended fuels. Therefore, part of the approval process should include a review of manufacturer data sheets or listings to confirm the component's compatibility with ethanol (Figure 12-22). **[Ref. 2306.8.2]**

LIQUEFIED PETROLEUM GAS DISPENSING

Liquefied petroleum gas (LP-gas) is a flammable liquefied compressed gas used as a motor vehicle fuel. In comparison to conventional fuels like unleaded gasoline and diesel fuel, LP-gas is the third most popular fuel used in the United States. Being a liquefied compressed gas, LP-gas is stored and dispensed as a liquid from a stationary container or tank into the vehicle fuel tank by way of a pump and piping. Before it can be used as a motor fuel, the vehicle must vaporize the liquid and convert it to a gas for carburetion.

LP-gas storage and dispensing must comply with the requirements in IFC Chapters 23, 53 and 61 and the requirements in NFPA 58, *Liquefied Petroleum Gas Code*. The stationary LP-gas container and its piping are sited and installed in accordance with NFPA 58 and Chapter 61. The dispenser, hose and nozzle are installed in accordance with all of the requirements in Chapter 23 and the NFPA 58 requirements for dispensers. **[Ref. 2307.5]**

Listed hoses, hose connectors including breakaway connections, vehicle fuel connectors (nozzles), dispensers, LP-gas pumps and electrical equipment are required (Figure 12-23). The storage container, piping, pressure relief devices and pressure regulators must be approved by the fire code official. **[Ref. 2307.2.1 , 2307.2.2]**

FIGURE 12-23 Listed LP-gas pump

FIGURE 12-24 The point of transfer is measured from the location where connections are made or disconnected.

The IFC requires that the point of transfer be separated from buildings, property line, streets and public ways. Point of transfer is the location where dispensing occurs: this is where the connection between the vehicle and the dispensing nozzle is made and broken (Figure 12-24). During any LP-gas dispensing operation, a very small volume of liquid is trapped between the nozzle and vehicle fill connection. Disconnection of the nozzle will cause this liquid to vaporize and release into the atmosphere. Therefore, the separation distances should be measured from the dispensing location. A minimum separation of 25 feet is required between the point of transfer and buildings with combustible exterior walls or non-combustible walls that have less than a 1-hour fire-resistance rating, lot lines on property that can be built on, streets, sidewalks and railroads. The point of transfer separation is reduced to 10 feet when buildings have exterior walls with a fire-resistance rating of 1 hour or more (Figure 12-25). [Ref. 2307.4]

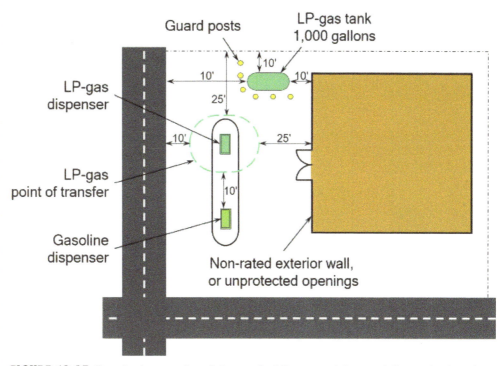

FIGURE 12-25 Required separation distances for LP-gas container and dispensing locations

NFPA 58 and the IFC set forth the requirements for the installation of LP-gas dispensers. An excess flow control valve and a manual shutoff valve are required in the piping between the transfer pump and the dispenser inlet. An excess flow control valve is designed to stop the flow of liquid LP-gas automatically in the event a pressurized pipe fails. The valve is sized based on the diameter of the pipe and the liquid flow rate. Therefore, if the diameter of the piping changes as part of the design or installation, an excess flow control valve is required at the point where the pipe diameter changes. Excess flow control valves are not required by the IFC or NFPA 58 to be listed (Figure 12-26). [Ref. 2307.5.1]

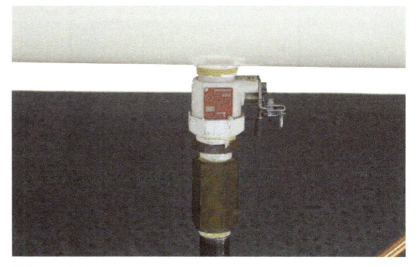

FIGURE 12-26 An internal excess flow control valve. Because of the size of the LP-gas container, it is equipped with a fusible link that closes the valve in the event of a fire.

A second excess flow control valve is required at the connection of the dispenser hose to the liquid LP-gas piping. This valve prevents the release of liquid LP-gas if the dispensing hose ruptures. An excess flow control valve is not required when a differential backpressure valve is used (Figure 12-27). A differential backpressure valve operates on the concept of constant flow and pressure. In the event of a

FIGURE 12-27 A differential backpressure valve

FIGURE 12-28 A listed dispenser nozzle for LP-gas

FIGURE 12-29 Hydrostatic pressure-relief valves

pipe or hose failure, the flow rate increases, which in turn causes the valve to close because of the pressure increase between the dispenser and tank. **[Ref. 2307.6.1]**

Dispenser hoses are listed for LP-gas service and equipped with a listed automatic-closing type nozzle valve. This valve is designed such that liquid LP-gas cannot flow until a positive mechanical connection is made at the vehicle fill connection (Figure 12-28). The installation of a listed emergency breakaway connection is required in the dispenser hose. **[Ref. 2307.5.1, 2307.5.3]**

All liquefied compressed gases exhibit fairly high expansion ratios at normal temperatures. Because LP-gas is a mixture of ethane, propane and butane, the expansion can vary, but it is generally about 36.4 cubic feet/gallon. At 70°F, LP-gas has a vapor pressure of 145 PSIG. Therefore, if the liquid is trapped in a closed pipe or hose, the pressure can increase as the pipe or hose is warmed to a point where catastrophic failure can occur. To prevent such a failure, the IFC specifies that the system must be protected from mechanical damage by an approved method. This requires the installation of one or more hydrostatic pressure relief valves. This valve is designed to safely vent the LP-gas once the pressure reaches a preset point. When the pressure is relieved, the valve closes. A hydrostatic pressure relief valve is required by NFPA 58 in any section of pipe or hose where liquid LP-gas can be trapped (Figure 12-29). This includes the LP-gas dispenser hose because liquid can be trapped between the dispenser valve and the LP-gas pump. **[Ref. 2307.6.2]**

HYDROGEN DISPENSING

The Department of Energy is encouraging manufacturers, and the country, to use renewable energy. One of those renewable energy sources is hydrogen. Hydrogen is the lightest substance, and it is odorless, colorless and tasteless. See Table 12-2 for the properties of hydrogen gas compasred to other fuels.

TABLE 12-3 Properties of Hydrogen[7,8]

Property	Hydrogen	Gasoline	Natural Gas	LP-gas
Vapor density	0.0696	4.0	0.56	1.52
Boiling point	-423°F	°F	-260°F	-44°F
Lower flammable limit (LFL)	4.0%	1.4%	5.3%	2.1%
Upper flammable limit (UFL)	75.0%	7.6%	15%	10.1%
Flammable range	71%	6.2%	9.7%	8%
Minimum ignition energy (MIE)	0.02 mJ	0.2 mJ	0.29 mJ	0.26 mJ

mJ = millijoules

7. "Hydrogen Safety," Hydrogen Tools Portal, Pacific Northwest National Laboratory, www.h2tools.org.
8. "Hydrogen Compared with Other Fuels," Hydrogen Tools Portal, Pacific Northwest Laboratory, www.h2tools.org.

Hydrogen is a desirable fuel for the environment since it is renewable, and can produce zero emissions. However, as with any fuel, there are fire risks associated with hydrogen. Some of the characteristics of hydrogen make it hazardous, such as the large flammable range and the low energy level required for ignition. The flammable range of hydrogen is ten times that of gasoline, and the energy required to ignite hydrogen is one-tenth of the energy required to ignite gasoline.

Hydrogen is considerably lighter than air and rapidly rises and disperses when released. Hydrogen will diffuse with air at a rate 3.8 times faster than natural gas. When released into the air, hydrogen will rise at a speed of about 45 miles per hour. For these reasons, hydrogen storage and compression facilities are allowed on the canopy top over the fueling island. [Ref. 2309.1.5]

Hydrogen can be captured from natural gas, and often hydrogen will be generated on-site. Hydrogen can be generated, stored and dispensed indoors or outdoors. Outdoor installations require similar safety provisions to dispensing facilities for natural gas or gasoline. One additional requirement for hydrogen is that when the emergency shutdown switch is operated, the hydrogen generation process also shuts down. [Ref. 2309.3.1.5]

Indoor generation of hydrogen must be accomplished in a hydrogen fuel gas room if the quantity in the room does not exceed the maximum allowable quantity, or a Group H-2 for quantities above the maximum allowable quantity. The hydrogen fuel gas room is considered an incidental use and *International Building Code* Table 509 requires the room to be separated by 2-hour fire-resistance-rated construction in Groups A, E, I and R occupancies. In all other occupancies, the room can be separated by 1-hour fire-resistance-rated construction. The hydrogen fuel gas room is to be provided with gas detection, mechanical ventilation, standby power and alarms (Figure 12-30). [Ref. 5808.1]

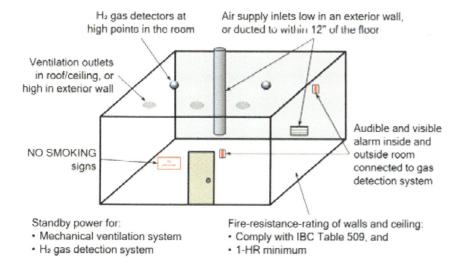

FIGURE 12-30 Safety features in a hydrogen fuel gas room

Code Essentials

"Mobile fueling" is fuel dispensing from a tank vehicle into multiple motor vehicles at a single location. Mobile fueling is also known as "fleet fueling." "On-demand mobile fueling" is dispensing from a tank vehicle, or other approved vehicle, into motor vehicles at multiple locations on an as-requested basis. •

ON-DEMAND MOBILE FUELING

A new process that is allowed in the code is designed to bring the fuel to the vehicle, rather than take the vehicle to the service station for fuel. Mobile fueling consists of a mobile fueling operator driving a fuel tank vehicle to the location where your car is parked. The mobile fuel operator will fill the fuel tank while you are at work, school or shopping. When you return to your vehicle, the fuel tank is full, and you are ready to be on your way.

In on-demand mobile fueling, the site is not as controlled as it is with a service station. The vehicle being fueled will typically be in a parking lot. To mitigate the hazards in these various locations, the IFC contains specific requirements that are necessary to conduct on-demand mobile fueling. [Ref. 5707]

On-demand mobile fueling operations require an operating permit. As part of the permit process, the fire code official will review the site where mobile fueling will occur. The operator is required to provide a detailed site plan of each location. On-demand mobile fueling must be at least 25 feet from buildings, property lines, combustible storage, storm drains and sources of ignition. The distance to storm drains can be reduced to 15 feet where the storm drain is covered, or precautions are taken to prevent a spill from entering the storm drain (Figure 12-31). [Ref. 105.6.16, 5707.4]

FIGURE 12-31 On-demand mobile fueling operations must be 25 feet from buildings, property lines, combustible storage, storm drains and any ignition source. *(Courtesy of Booster Fuels, Inc.)*

The on-demand mobile fueling vehicle must be equipped with a spill kit to handle a 5-gallon spill. The on-demand mobile fueling vehicle must either comply with NFPA 385, Standard for Tank Vehicles for Flammable and Combustible Liquids or carry no more than 60 gallons of fuel in approved metal safety cans or other approved containers. All vehicles must carry a 40-B:C fire extinguisher. [Ref. 5707.2, 5707.5]

A safety and emergency response plan is required which covers the fueling operations and safety parameters, spill prevention and staff training requirements. Mobile fueling operators must be trained in the dispensing procedures and proper mitigation of fuel spill. [Ref. 5707.3]

You Should Know

- The requirements in Chapters 23 and 57 are designed to allow safe operation and use of flammable liquids and flammable gases for handling and dispensing fuels. •

CHAPTER 13
Flammable Finishes

Unique processes and materials are used in the application of flammable finishes. Liquids can be atomized into an aerosol and sprayed onto objects, or goods can be immersed into open tanks. In a process known as powder coating, ionized solids are suspended in air and attach to metal objects that have been given an electrical charge. Rigid thermoplastics are manufactured using glass fibers that are applied using a resin and chemically activating it using a chemical catalyst. Chapter 24 of the *International Fire Code* (IFC) contains requirements that address processes where flammable and combustible liquids or combustible dusts are applied onto metal, wood or plastic goods to provide aesthetically pleasing and durable coatings.

You Should Know

The title "Flammable Finishes" does not mean that Chapter 24 only applies to finish applications using "flammable liquids." IFC Chapter 24 covers finishing operations using combustible liquids, combustible powders, dip tanks and fiberglass in addition to flammable liquids.

Section 2403 contains the minimum requirements for the protection of operations involved in the application of flammable finishes. This section addresses sources of ignition, storage, use and handling of flammable and combustible liquids and the operation and maintenance of flammable finishing activities. Section 2403 is applicable to all of the activities regulated in Sections 2404 through 2410.

TYPES OF FLAMMABLE FINISHING PROCESSES

Paints and coatings applied by spray finishing can be hydrocarbon solvents or water based. They are formulated with film-forming chemicals and pigments. Properly applied film forming chemicals bond together and create a seamless surface. The pigment is the color in the paint or coating. Generally, a solvent (either hydrocarbon or water) is added to the coating to carry the film forming chemicals and pigments to the surface onto which they are being applied. A catalyst may be added to the mixture to initiate the polymerization of the film forming process and accelerate drying. In many industries, solvent-based paints and coatings are preferred for finish quality reasons and they generally dry quicker when compared to water-based paints. Hydrocarbon solvent products may release volatile organic compounds (VOCs) that are a fire and environmental hazard.

FIGURE 13-1 Spray finishing operation in a spray booth

Spray finishing with flammable or combustible liquids is a process where the liquid is pressurized and discharged through a very small orifice to create an aerosol (Figure 13-1). Aerosols are very small droplets of flammable or combustible liquids suspended in air. The droplets have very little mass but a large surface area. The droplets in air are easily ignited because they quickly evaporate into vapor. By discharging a large number of very small flammable or combustible liquid droplets under pressure, spray finishing produces a large volume of flammable vapor.

Powder coating is the application of thermoforming or thermosetting plastic powders onto a metal surface (Figure 13-2). Common powders include polyester, polyester-epoxy or acrylics. The powders range in size from 20 to 40 microns, which is comparable to the size of baking flour. Powder coating does not require a solvent to bond the powder onto the metal; therefore, it produces less pollution and essentially no hazardous waste when compared to petroleum solvent-based liquids.

FIGURE 12-2 Powder-coating operation

After the surface is cleaned, the powder is applied onto the metal surface. The powder receives a positive electrical charge from the powder coating gun or nozzle.

The charged particles adhere to metal surfaces, which are negatively charged. The oppositely charged powder is attracted and adheres to the metal parts. Powder coating guns generally are used for manual application while powder coating nozzles are found on automatic or robotic equipment.

Suspension of the powder in air represents the greatest hazard. The powder is a combustible dust and can burn or contribute to a deflagration. A deflagration can occur when the concentration of the powder is within its minimum explosive concentration (MEC) and is suspended in air, a strong enough source of ignition is present and the dust is confined inside of a powder coating booth or room. The potential for dust deflagration extends to dust collection equipment, especially powder coating operations using remote dust collectors that are connected to the powder coating equipment using ducts. If during the spray operation the spray gun is moved too close to the object receiving the powder, a spark could jump across the air gap between the negative-charged surface and the positive-charged spray gun. To eliminate this occurrence, several specific safeguards are required:

- the electrical charge produced by the spray gun must be of less intensity than is required to ignite the combustible powder even above the MEC,
- the spray booth and all electrically conductive equipment must be properly grounded and
- adequate ventilation is provided to minimize the dust cloud created in the spray booth.

Manufacturing of reinforced plastics involves the use of a glass-fiber mat (also known as "rove"), which is placed onto a mold of the desired shape. The glass-fiber mat is coated or sprayed with resin that is mixed with a catalyst to accelerate the drying rate of the resin. Reinforced plastics are found in the construction of boats, plumbing fixtures such as sinks and bathtubs, and many automotive applications.

The most common resin that is used is unsaturated polyester resin (UPR). The primary constituent of UPR is styrene monomer. Most UPRs are classified as either Class IC flammable or Class II combustible liquids and may also be classified as either a Class 1 or Class 2 unstable (reactive) liquid.

A catalyst is added to promote drying and the uniform hardening of the resin. The catalyst most commonly used is the organic peroxide methyl ethyl ketone peroxide (Figure 13-3). Organic peroxides are a unique class of hazardous materials because they contain hydrocarbon molecules connected or branched to oxygen molecules. In other words, there is a fuel (hydrocarbon) and a source of oxygen combined into the one material. Organic peroxides are sensitive to any contamination and also have a limited shelf life based on the storage temperature

FIGURE 13-3 Methyl ethyl ketone peroxide is an organic peroxide that can be easily ignited and has a limited shelf life.

During the manufacturing process of reinforced plastics, the resin and the catalyst will generally be mixed together manually or at the spraying or "chopper" gun (Figure 13-4). A chopper gun is a pneumatically-powered, hand-held or robotically-controlled device that simultaneously applies glass fiber strands, resin and the catalyst. The process can involve the manual smoothing of the resin, fibers and glass fibers using paint rollers.

FIGURE 13-4 Chopper gun used in the manufacturing of reinforced plastics

SPRAY BOOTH AND SPRAY ROOM CONSTRUCTION

IFC Section 2404 specifies requirements for the construction of spray spaces, booths and rooms and the spray application of flammable finishes. IFC Section 2404 requirements stipulate the location of spray finishing operations, the design and construction of spray rooms, spray spaces and spray booths, fire protection, control of ignition sources, mechanical ventilation systems, interlocks required for spray finishing and limited spray spaces. Section 2404.1 requires that spray finishing activities also comply with the requirements in Section 2403 for the protection of operations.

Spray finishing is allowed in a spraying space, spray booth or spray room. In Group A, E, I, and R occupancies, spray finishing can only occur in a spray room that is separated by fire-resistance-rated construction in accordance with the *International Building Code* and is protected by an automatic sprinkler system designed in accordance with Section 903.3.1.1. In all other occupancies, spray finishing can be performed in a spray booth or an approved spraying space. [Ref. 2404.2]

Whether the operation occurs in a spray room, spray booth or spraying space, the IFC prescribes specific safety features to protect the opera- tors and the building. In general, the safety features are identified in Table 13-1.

In some occupancies, the code official may approve limited spraying spaces. Limited spraying spaces commonly are found in automobile body shops where only small jobs are performed. In order to qualify as a limited spraying space

- The aggregate surface area to be sprayed shall not exceed 9 square feet. This is large enough to allow the painting of hoods or trunk lids.
- Spraying operations cannot be continuous.
- Positive mechanical ventilation providing a minimum of six complete air changes per hour is required and must meet the *International Mechanical Code* (IMC) requirements for hazardous exhaust systems.

- Electrical wiring within 10 feet of the floor and 20 feet horizontally of the limited spraying space shall be designed for Class I, Division 2, locations in accordance with NFPA 70, *National Electrical Code®*. **[Ref. 2404.9]**

TABLE 13-1 Safety features in spray rooms, spray booths and spraying spaces

Safety feature	Spray room	Spray booth	Spraying space
Occupancy classification	F-1 or H-2 depending on quantities	NA—considered an appliance	NA—part of approved building operations
Fire-resistance-rated separation to other portions of the building	✓	NR	NR
Interior surfaces must be noncombustible, smooth and designed to withstand frequent cleaning	✓	✓	NR
Width of exit doors	36 inches	30 inches	As required by Chapter 10
Floor shall be noncombustible or covered with noncombustible material	✓	✓	✓
Size limitation	IBC Table 506.2	1500 square feet	NA
Automatic fire-extinguishing system	✓	✓	✓
Mechanical ventilation	Must maintain exhaust < 25% LFL[a]	100 feet/minute for open face spray booth Must maintain exhaust < 25% LFL[a]	6 air changes/hour
Spray equipment interlocked with ventilation equipment	✓	✓	NR
Electrical – Class I, Division 2	Inside the spray room	Inside the spray booth and within 3 feet of doors or open face	Within 20 feet horizontally and 10 feet vertically
Control of ignition sources	✓	✓	✓

NA = Not Applicable
NR = Not Required
a. Spray booths and spray rooms with fixed or electrostatic spray application must provide 50 feet/minute through all openings.

A spray room is an actual room constructed as part of the building. IBC Table 509 requires that spray rooms be separated by a minimum of 2-hour fire-resistance-rated construction from the remainder of the building, or if the building is sprinklered the separation can be reduced to 1-hour fire-resistance-rated construction. A spray booth, on the other hand, is considered an appliance that is placed into the building (Figure 13-5). Therefore, a spray booth does not receive an occupancy classification, and the separation requirements to provide an acceptable level of fire resistance for a booth are not the same as for a spray room. The fire resistance of the spray booth is based on noncombustible construction of 18-gauge steel along with a clear space of 3 feet around the spray booth. Since this is not equivalent to the 1-hour fire-resistance rating of a spray room, spray booths are limited in

FIGURE 13-5 A downdraft spray booth inside a building

maximum size. For example, a spray room classified as Group H-2 has a tabular area of 7,000 square feet when constructed of noncombustible materials. A spray booth constructed of noncombustible materials is limited to 1,500 square feet. The IFC and IBC regulate the construction of the spray room, while a spray booth must meet the requirements of the IFC and NFPA 33, Standard for Spray Application Using Flammable or Combustible Materials. **[Ref. 2404.3, IBC 416.2, IBC Table 506.2]**

All spray rooms and booths require mechanical ventilation. A properly designed and maintained mechanical ventilation system provides sufficient supply and exhaust air to maintain the atmosphere inside of a spray room or booth below 25 percent of the lower flammable limit of the most volatile flammable liquids sprayed. The mechanical ventilation system dilutes the flammable vapor with air by mixing it inside of a plenum and safely discharging the vapor/air mixture outdoors or into a pollution control device. Dry or wet filters are provided to capture pigments and film formers and prevent them from being conveyed through the mechanical ventilation system. **[Ref. 2404.7.3]**

Spray rooms and booths are constructed of approved noncombustible materials. The IFC requires the interior wall surfaces and any surfaces where accumulations can be deposited to be smooth in construction so as to allow free air movement and to facilitate cleaning. The IFC prohibits the use of aluminum in the construction of spray rooms and spray booths. Aluminum exhibits a much lower melting point when compared to carbon steel—aluminum would limit a booth's ability to confine a fire. **[Ref. 2404.3.1, 2404.3.3]**

Walls, floors and ceilings, as well as their exhaust ducts and all associated components that form the spray booth appliance, are required to be constructed of noncombustible materials. In most cases, minimum 18-gage sheet metal is used for spray booth construction. However, some manufacturers will create a structural panel using a wall assembly fabricated with a structural steel channel that is covered on both sides by 20-gage steel. Spray booths constructed using sheet metal wall and ceiling components are shipped unassembled and erected on site. To ensure that the airflow remains uniform during the operation of spray booths, the IFC allows wall and ceiling joints to be sealed by caulks or other similar sealants (Figure 13-6). **[Ref. 2404.3.3.1]**

FIGURE 13-6 A downdraft spray booth constructed of approved, noncombustible materials

MECHANICAL VENTILATION

Of all of the safety features installed in a spray booth or room, mechanical ventilation has the greatest positive impact on the protection of the booth, its occupants (when occupied) and the goods being finished. A properly designed and maintained mechanical ventilation system safely removes flammable vapors or combustible dusts and exhausts them to a safe location. The mechanical ventilation system maintains the atmosphere inside of the flammable vapor area below 25 percent of the lower flammable limit for the most volatile flammable liquid or 50 percent of the minimum explosive concentration for combustible dusts (Figure 13-7). The design of these systems must also comply with the hazardous exhaust system requirements in Section 510 of the *International Mechanical Code*. [Ref. 2404.7.3]

FIGURE 13-7 The design of these systems must also comply with the hazardous exhaust system requirements in Section 510 of the *International Mechanical Code*.

When compared to other mechanical ventilation systems, hazardous exhaust systems have unique design requirements:

- The ventilation system design must use uncontaminated supply air to dilute and maintain the flammable constituents at concentrations below 25 percent of its lower flammable limit.
- The exhaust system, including the exhaust duct, fans and fan motors, must be independent of other building mechanical exhaust systems. With exceptions for laboratories, hazardous exhaust systems shall not share common shafts with other air-handling ducts.
- A hazardous exhaust system for a flammable vapor area must be designed using the constant velocity method.
- The minimum airflow velocity into the spray room or booth with opening is 100 feet/minute through all openings into the spray booth. This can be reduced to 50 feet/minute where an electrostatic spraying process is utilized.
- The termination point of the duct system must be at least 30 feet from property lines and 10 feet from operable building openings. [Ref. 2404.7.3, IMC 510]

A condition of acceptance is an airflow balance testing. In flammable vapor areas, balancing is accomplished by ensuring the volume of makeup air is approximately equal to the rate that air is exhausted. One method of verifying this is requiring an air balance test report, which documents the measurements of the air velocity in the supply air source and the exhaust duct to ensure the exhaust fan is discharging at its design capacity. For example, if a spraying area is required to be protected by an exhaust fan rated at 15,000 cubic feet/minute at a given static air pressure, the air balance test report will document if the fan is performing as specified. Air balance test reports can be prepared by a booth manufacturer's representative or heating, ventilating and air-conditioning firms (Figure 13-8).

FIGURE 13-8 An air balance test can accurately and economically confirm that the hazardous exhaust system complies with the IFC and IMC.

Because the mechanical ventilation system for a spraying area must comply with IMC Section 510, independent exhaust ducts are required from each spray booth, room or area (Figure 13-9). The IFC requires that the exhaust duct meet the following certain separation distances at the point of discharge:

- Property lines: 30 feet
- Operable building openings: 10 feet
- Height above exterior walls and roofs: 6 feet
- Combustible walls or openings into the building that are in the direction of the exhaust discharge: 30 feet.
- Above adjoining grade: 10 feet. [Ref. 2404.7.6]

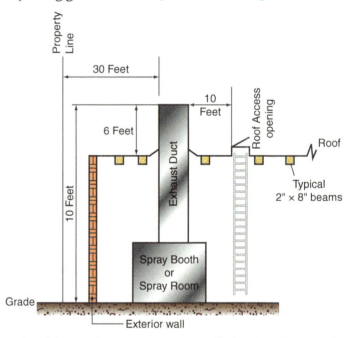

FIGURE 13-9 Separation distances for the termination point of the ventilation system are specified in the IFC and IMC.

To ensure that the velocity of the exhaust air meets the required flow rate, the IFC requires a means of measuring airflow be provided using a visual gauge, pressure switch or an audible alarm to indicate a reduction in the required airflow velocity. The most common method used is the installation of a manometer that measures the air pressure at the intake and exhaust side of the booth's exhaust plenum. The manometer measures the differential pressure of the exhaust air. A common unit of measurement is inches of water column. (27.71 inches of water equals one pound/square inch gauge).

As the filter becomes loaded with film forming chemicals and pigments, the differential pressure increases. The manometer measuring tube on the intake side (the spraying area) will register the increased pressure that results from the increased resistance to airflow caused by the filter debris. When the differential pressure exceeds the manufacturer's pressure limits, the filters should be replaced (Figure 13-10). [Ref. 2404.7.8.3]

FIGURE 13-10 Differential pressure manometer *(Courtesy of Dwyer Instruments Inc., Michigan City, IN)*

ILLUMINATION

To facilitate the application of flammable finishes, the interior of spraying areas is illuminated by electrical lamps. The use of electric lamps creates a potential ignition source because of the heat they produce or when lamps are damaged, their energized circuits can be exposed. The IFC prohibits portable electric lamps for the illumination of flammable vapor areas and requires that when portable lamps are used, they must be limited to cleaning or repair operations and approved for hazardous (classified) electrical atmospheres. [Ref. 2404.6.2.4]

Fixed lamps or luminaires illuminate the interior of a spraying area. The requirement for lamps depends on their design. Luminaires are either external or internal (termed as "integral" in the IFC). External luminaires are designed where the light bulbs are in an area that is not classified as a hazardous location. Internal (integral) luminaires are designed where the light bulbs are located and replaced in an area within a hazardous (classified) location. Hazardous (classified) locations are areas where flammable vapors are or may be present. Electrical equipment installed in hazardous (classified) locations must be listed for use in these atmospheres so it does not become an ignition source. [Ref. 2404.6.2]

External luminaires are attached to the walls or the ceiling of a spraying area (Figure 13-11). The luminaires are designed for the replacement of the lamps outside of the spray booth so as to maintain the ordinary hazard electrical classification. [Ref. 2404.6.2.2]

Integral luminaires are designed and listed for repair inside of a flammable vapor area (Figure 13-12). The fire code requires lamps listed for Class I, Division 2, locations for flammable vapor areas and Class II, Division 2, locations when the luminaire is used for the illumination of powder coating areas. The luminaire also must be evaluated and listed to ensure that residual deposits cannot be heated and ignited by radiant energy from the lamps. [Ref. 2404.6.2.3]

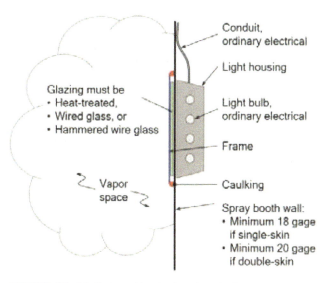

FIGURE 13-11 External luminaire electrical components are located outside of the boundary of the hazardous (classified) area.

FIGURE 13-12 Integral luminaire *(Courtesy of Cooper Industries—Crouse Hinds Electrical Division, Syracuse, NY)*

FIGURE 13-13 An electric solenoid valve installed in compressed air piping

INTERLOCKS

A properly designed mechanical ventilation system includes an interlock with the spraying equipment. The interlock is arranged so the spraying apparatus will not operate if the mechanical ventilation system is not operating. [Ref. 2404.7.1]

There are a number of methods that may be used to satisfy this provision. The most common method is to install an electric solenoid valve in the compressed air piping supplying the spraying equipment (Figure 13-13). The solenoid valve should be located so it is not within the hazardous (classified) electrical boundary, although valves are available that are listed for use in Class I, Division 2, areas. The solenoid valve is wired into the mechanical ventilation fan motor. When the motor is energized the solenoid valve opens, allowing the flow of compressed air to the spraying equipment.

Another method for satisfying the interlock provision is installing a differential pressure switch that is designed to measure pressure within two areas, such as the spray space and the exhaust plenum (Figure 13-14). If the pressure difference exceeds a given value—generally measured in inches of water column—the pressure switch operates and allows the spray equipment to operate. The pressure differential switch is considered a more reliable method of interlocking spray equipment with the mechanical ventilation system because the fan will not operate if the paint filters are excessively loaded with overspray. Many spray booth fans are driven by belt drives and a fan may require two or three belts. If a fan belt breaks, the differential pressure will be lower than the operating range of the pressure switch, preventing the flow of compressed air to the spraying equipment.

FIGURE 13-14 Differential pressure switch

FIRE PROTECTION

Spray booths and spraying rooms are required to be equipped with an automatic fire-extinguishing system. The IFC requires that the fire-extinguishing system area of protection includes exhaust plenums, exhaust ducts and both sides of dry filters when they are used. [Ref. 2404.4]

NFPA 33 states that a spray booth or spray room can be protected using any of the following fire-extinguishing systems:
- Automatic sprinkler system – NFPA 13, Standard for the Installation of Sprinkler Systems
- Automatic foam-extinguishing system – NFPA 16, Standard for the Installation of Foam-Water Sprinkler and Foam-Water Spray Systems

- Carbon dioxide extinguishing system – NFPA 12, Standard Carbon Dioxide Extinguishing Systems
- Dry-chemical extinguishing system – NFPA 17, Standard Dry Chemical Extinguishing Systems
- Clean-agent extinguishing system – NFPA 2001, Standard Clean Agent Fire-Extinguishing Systems
- Automatic water mist system – NFPA 750, Standard Water Mist Fire Protection Systems.

When an automatic sprinkler system is used, NFPA 33 requires the system be designed for an Ordinary Hazard Group 2 discharge density for powder coating and styrene thermoset resin applications and an Extra Hazard Group 2 discharge density for all other spray operations. NFPA 33 also requires the water supply for the sprinklers installed in the spray booth be controlled by a separate, listed control valve that is accessible from the floor level. NFPA 33 contains specific requirements for the installation of automatic sprinklers in the exhaust ducts and stacks, including maximum spacing requirements and minimum discharge flow rates.

Dry chemical alternative fire-extinguishing systems are commonly used for the protection of spray booths or rooms. For spray rooms in a sprinklered building, the automatic sprinkler system must be provided as the fire-extinguishing system in the spray room in order for the building to be considered sprinklered throughout. However, for a spray booth in a sprinklered building, a dry-chemical fire-extinguishing system may be used to satisfy the fire protection requirement of the spray booth. In this case, the building would be considered sprinklered throughout since the spray booth is an appliance rather than part of the building. If the building is not equipped with an automatic sprinkler system, then either a fire sprinkler system or a dry-chemical fire-extinguishing system can be installed to protect the spray room or spray booth. In all situations, a dry-chemical fire-extinguishing system must be listed for protection of a spray room or booth. The design of dry chemical extinguishing systems must comply with NFPA 17, Standard for Dry Chemical Extinguishing Systems. [Ref. 904.6]

NFPA 17 permits the use of engineered or pre-engineered dry chemical fire-extinguishing systems (Figure 13-15). Total flooding systems are specified for the protection of spraying areas. A total flooding system is designed to discharge the fire-extinguishing agent into an enclosure that surrounds the hazard. In a spray booth or room, the interior of the spray space is the hazard to be protected. Fire protection system manufacturers offer a variety of pre-engineered total flooding designs for the protection of open-face, cross-draft and down-draft spray booths.

FIGURE 13-15 Pre-engineered dry-chemical fire-extinguishing system protecting a spray booth *(Courtesy of TYCO/Ansul Inc., Marinette, WI)*

CHAPTER 14

High-Piled Combustible Storage

With the cost of land increasing and changes in technologies that allow for rapid fulfillment of a customer's request for manufacturing and consumer goods, high-piled combustible storage is a preferred method of handling product in storage (Group S) and many mercantile (Group M) occupancies. Many retailers for home improvement, consumer electronics and consumer soft goods such as furniture, clothing, food and beverages construct large Group M occupancies that use high-piled combustible storage to satisfy their customer demands and to control costs.

High-piled combustible storage methods allow for a greater amount of materials and product to be stored within a given floor area. This increases the potential fire dollar loss for each square foot of floor area, with property losses resulting from smoke and fire damage of goods within the building in many cases often surpassing the construction cost of the building.

Possibly the most important fire protection in a high-piled combustible storage building is the automatic sprinkler system. In the past 15 years, the fire protection engineering community has witnessed the introduction of a variety of new automatic sprinklers that are designed specifically for high-piled combustible storage. As a result, it can be challenging to verify the adequacy of the automatic sprinkler system for the product stored. The available methods of design have also increased and NFPA 13, Standard for the Installation of Sprinkler Systems, requirements for automatic sprinkler protection continue to be extensively modified.

WHAT IS HIGH-PILED COMBUSTIBLE STORAGE?

High-piled combustible storage is the "storage of combustible materials in closely packed piles or combustible materials on pallets, in racks or on shelves where the top of storage is greater than 12 feet in height. When required by the fire code official, high-piled combustible storage also includes certain high-hazard commodities, such as rubber tires, Group A plastics, flammable liquids, idle pallets and similar commodities, where the top of storage is greater than 6 feet in height." [Ref. 202]

The provisions in Chapter 32 of the *International Fire Code* (IFC) become applicable when the product stored, the storage containers or the pallets are combustible and the storage is in piles, on pallets in racks, or on shelves and the height of storage is greater than 12 feet. The 12-foot limit is based on fire testing that was performed in the 1960s and 1970s. The tests revealed that when goods were stored more than 12 feet in height, the rate of fire spread and heat release dramatically increased to the point that a sprinkler system with a typical ordinary hazard design for a retail establishment in accordance with NFPA 13 could not control the fire.

The threshold for application of Chapter 32 for commodities classified as high-hazard is 6 feet. They include rubber tires, highly combustible and fast-burning plastics (termed "Group A"), flammable and combustible liquids, idle pallets, and alcohol or hydrocarbon formulated aerosols. NFPA 13, Chapters 15 through 20 contain automatic sprinkler design criteria for storage of high-hazard commodities including rubber tires, Group A plastics, candles, rolled paper, wax-coated paper cups and idle wood or plastic pallets. High-hazard commodities almost always require a specialized automatic sprinkler system that is designed to control or suppress fires involving these goods (Figure 14-1). [Ref. 903.3.1.1]

FIGURE 14-1 Flammable and combustible liquids in plastic containers represent a significant challenge in the design of automatic sprinkler systems and are treated as high-hazard commodities by the IFC.

Code Essentials

The "high-piled storage area" is defined as an area within a building that is designated, intended, proposed or actually used for high-piled combustible storage. ●

COMMODITY CLASSIFICATION

To determine if the requirements of IFC Chapter 32 are applicable, the stored commodities must be classified in accordance with Section 3203, the height of storage must be established and the high-piled combustible storage area be designated. The commodity classification and the height of storage will influence the design of the automatic sprinkler system when the system is required.

Commodities are classified based on an estimation and comparison of the heat release rate of typical products in the category. The overall fire hazard of a commodity is a function of its heat release rate. Heat release rate is measured in BTU/minute (kW). The heat release rate is the product of the heat of combustion measured in BTU/pound (kJ/kG), and the burning rate is measured in pounds/minute (kg/second). The higher a commodity's heat of combustion and heat release rate, the higher the commodity classification. The higher a commodity's classification, the more difficult this material is to control and extinguish when it is involved in fire.

The IFC separates hazards into the following five commodity classifications:

- Class I
- Class II
- Class III
- Class IV
- High-hazard.

Figure 14-2 illustrates that a Class I commodity is considered to have the lowest heat release rate, while a high-hazard commodity has the highest heat release rate.

High-hazard Commodity	Highest Fire Hazard
Class IV Commodity	↑
Class III Commodity	
Class II Commodity	
Class I Commodity	Lowest Fire Hazard

FIGURE 14-2 The higher the commodity class, the greater the challenge in the design of the automatic sprinkler system.

Before making a determination of a commodity's hazard classification, it is important to review the definition of "commodity," which is "a combination of products, packing materials and containers." Flammability properties of the product, packaging and containers need to be evaluated in establishing appropriate fire protection. The commodity classification is not limited to the packaged product but also includes the packing material and the material of construction for the container. Packing materials can include a large amount or volume of plastics, which generally exhibit higher heat of combustion when compared to other materials constructed of noncombustible and combustible materials. Consider the definition of a Class I commodity and compare it to a Class II commodity. Class I commodities are essentially noncombustible materials and products contained within combustible packaging or stored on combustible pallets. For example, ceramic coffee mugs packaged in a single-thickness cardboard box are Class I commodities (Figure 14-3). A Class II commodity

FIGURE 14-3 Single-thickness cardboard

includes Class I products that are packaged in slatted wooden crates, solid wooden boxes, multiple layer paper or fiberboard cartons, or equivalent combustible packaging materials with or without pallets. In other words, the packaging material is changed resulting in a higher commodity classification. The ceramic coffee mugs in a multiple-thickness cardboard box with cardboard dividers become a Class II commodity (Figure 14-4). [Ref. 202, 3203.2, 3203.3]

When classifying a commodity, the pallet itself is not considered as part of the commodity, and is intentionally not included in the definition of commodity. The product on the pallet is classified and then the pallet is added. The commodity classification and the sprinkler design assume there is a wooden pallet. So, when the pallet is not there, the product is still adequately protected. See the discussion on Page 201 regarding the use of plastic pallets, which can be different than wooden pallets (Fig 14-5).

NFPA 13, Standard for the Installation of Sprinkler Systems, includes only the commodity classifications of Class I through IV. The commodities that fall into the IFC classification of "high-hazard" fall under special requirements in NFPA 13 or other NFPA standards. For example, the IFC includes rubber tires, rolled paper and aerosols as high-hazard commodities. NFPA 13 addresses requirements for rubber tires and rolled paper in Chapters 18 and 19 respectively, but refers aerosols to NFPA 30B, *Code for the Manufacture and Storage of Aerosol Products*. The fact that NFPA 13 does not include the term high-hazard has no effect on the application of the automatic sprinkler system design. The code user will end up at the same location for the appropriate design criteria. However, the classification of high-hazard does change requirements in the IFC, such as smoke and heat removal, allowable height of storage, allowable size of high-piled storage area and thresholds for automatic fire sprinklers.

It is important that the physical form of the commodity be evaluated when assigning the commodity classification. The physical form of a commodity is commonly referred to as its geometry, which must be considered when classifying commodities. Materials will burn differently based on their geometry. For example, consider dimensional wood lumber. Because of its compacted mass, it is difficult to burn when stored horizontally on a pallet load. However, if the same dimensional lumber is cut and assembled to create pallets, more exposed surface area is created. The result is that a material that is normally classified as a Class III commodity by Table 3203.8 (listed as "wood products—lumber") is now classified as a high-

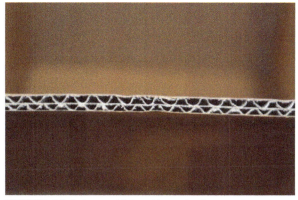

FIGURE 14-4 The additional layer in the cardboard is referred to as a multiple-thickness cardboard and it increases the commodity classification.

FIGURE 14-5 This product is tomatoes and tomato paste in metal cans, with a paper wrapper, set in a cardboard box. This is a Class I commodity, and is still a Class I commodity when placed on a wooden pallet..

Code Essentials

The term "commodity" includes
- the product,
- the packaging material,
- the container
- and possibly shrink wrap used to retain the stored items on the pallet.

FIGURE 14-6 The geometry influences the commodity classification. The dimensional wood on the left is a Class III commodity, while the finished wood fence sections on the right are a high-hazard commodity.

hazard commodity by Table 3203.8 (listed as "miscellaneous—pallets and flats that are idle; combustible") simply by changing the geometry of the wood (Figure 14-6). **[Ref. 3203.8]**

A review of the Class III, IV and high-hazard commodity classifications finds that they include plastics. Plastics are classified separately in the IFC and NFPA 13 because they have such a wide range of heat of combustion per unit of mass that must be considered when assigning classifications. Because of the large number of plastics, the complexity of their nomenclature and the ease of changing burning characteristics with additives, careful consideration of the resins is necessary when classifying plastics. Certain plastics pose a significantly greater hazard than ordinary combustibles; therefore, plastics are classified separately.

The heat release rate (Btu/min or kW) for plastics can be two to three times greater than for a similar arrangement of ordinary combustibles. For example, the heat of combustion of ordinary combustibles, such as wood or paper, generally ranges between 6,000 and 8,000 Btu/lb. The heat of combustion for plastics generally ranges between 12,000 and 20,000 Btu/lb. The burning rate of a commodity is dependent on many things, but plastic materials generally exhibit higher maximum burning rates than similarly arranged ordinary combustibles.

Plastics are classified as either Group A, Group B or Group C. Group A plastics represent the most challenging classification from a fire protection viewpoint, while Group C plastics are the least challenging. **[Ref. 3203.7]**

The geometry of the plastic is important, as it will influence the ease of ignition and the burning rate of plastics. Plastics have three basic geometric forms:

- Expanded
- Unexpanded
- Free-flowing.

Expanded plastics are generally a low-density product and are commonly called "foamed plastics" such as polystyrene foam coffee cups or packaging material, and polyethylene and polypropylene foam sheeting. Expanded plastics have a cellular structure made up of many small air pockets and voids. The pockets increase the available surface area and promote easy ignition and increased vertical flame spread (Figure 14-7).

Unexpanded plastics have a higher density when compared to expanded plastics and may or may not

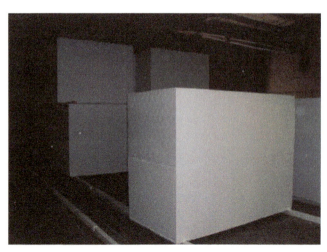

FIGURE 14-7 Expanded Group A plastic

have rigidity. Plastic films are classified as unexpanded plastic as well as plastic sheets. They are a solid material and can include goods such as toys, tote bins and containers. In comparison to expanded plastics, unexpanded plastics are less hazardous because of the reduced surface area (Figure 14-8).

Free-flowing plastics are very small plastic items such as bottle caps, hypodermic needle plungers, granular or flake plastics or powdered plastics. Free-flowing plastics burn less severely when compared to expanded or unexpanded plastics. When a package of free-flowing plastics fails under fire exposure, the products spill from the container. In rack storage, the items begin to fill the rack flue spaces and start to smother the fire slowing the fire spread. Because of this phenomenon, free-flowing plastics are classified as a Class IV commodity (Figure 14-9). [Ref. 3203.5]

Group A plastics are commonly used in packaging. Whether they are used in expanded plastic or unexpanded state, they are frequently seen and can easily be overlooked because the products they protect may often be noncombustible (Figure 14-10).

FIGURE 14-8 Unexpanded Group A plastic box

FIGURE 14-9 Free-flowing Group A plastic

FIGURE 14-10 This tomato box is constructed of polyethylene terephthalate, an unexpanded Group A plastic.

HIGH-PILED COMBUSTIBLE STORAGE AREAS

After the individual commodities are classified, the area of high-piled storage is designated. A high-piled storage area is "an area within a building that is designated, intended, proposed or actually used for high-piled combustible storage, including operating aisles." The size of the high-piled storage area will be used to determine the type of protection when applying Table 3206.2 requirements (Fig 14-11). [Ref. 202]

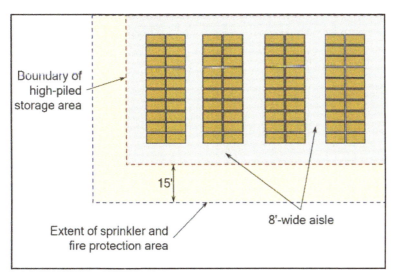

FIGURE 14-11 A high-piled storage area is the actual footprint of the storage arrays and includes any interior aisles and an aisle's width around the storage.

The IFC requires that the most challenging commodity be used as the basis for classification of the high-piled storage area; if a storage area has a mix of Class I, II, III and IV commodities, fire protection must be provided for the Class IV hazard level. This designation will not impact the design of fire department access and fire-fighter access doorways. It will, however, impact the design of the automatic sprinkler system and possibly the smoke and heat removal design. [Ref. 3204.1]

The Class IV commodity designation is the most commonly applied classification and protection application that is assigned to high-piled storage areas. Excluding high-hazard commodities, this assignment is common because it allows maximum flexibility in the proposed, current or future use of buildings. Many jurisdictions require the classification and protection of speculation warehouses based on Class IV commodities.

After the commodities are classified and storage height and high-piled storage area have been determined and identified, the amount and type of protection required for the facility must be determined. These requirements are set forth in Table 3206.2, which contains the general fire protection and life safety requirements for high-piled combustible storage. This table is an index to the requirements in IFC Sections 3206.2 through 3206.10.

A basic review of this table reveals that all of the requirements are dependent on the commodity classification and the size of the high-piled storage area. The table is divided into two basic commodity categories: 1) Classes I through IV and 2) high-hazard. For Class I through IV commodities, there are no real differences in the requirements in the IFC—once a classification has been assigned to one of four of these commodity classes, the requirements of Table 3206.2 are primarily based on the size of high-piled storage area. However, the automatic sprinkler system design criteria in NFPA 13 will change based on the Class I through Class IV designation.

Automatic sprinkler protection is required for high-piled combustible storage areas over 12,000 square feet housing Class I through IV commodities or storing high-hazard commodities in an area greater than 2,500 square feet. For instances where a particular storage area is designed for a lower hazard commodity and a higher hazard commodity is introduced, Section 3206.2 can be used to require improvements to the fire protection systems so the higher hazard commodity can be safely stored (see Table 14-1). [Ref. Table 3206.2]

TABLE 14-1 General fire protection and life safety requirements (IFC Table 3206.2)

Commodity class	Size of high-piled storage area[a] (square feet) (see Sections 3206.2 and 3206.3)	All storage areas (see Sections 3206, 3207 and 3208)[b]				Solid-piled storage, shelf storage and palletized storage (see Section 3207.3)		
		Automatic fire-extinguishing system (see Section 3206.4)	Fire detection system (see Section 3206.5)	Fire department access doors (see Section 3206.7)	Smoke and heat removal (see Section 3206.8)	Maximum pile dimension[c] (feet)	Maximum permissible storage height[d] (feet)	Maximum pile volume (cubic feet)
I–IV	0–500	Not Required[a]	Not Required	Not Required	Not Required	Not Required	Not Required	Not Required
	501–2,500	Not Required[a]	Yes[g]	Not Required	Not Required	120	40	100,000
	2,501–12,000 Open to the public	Yes	Not Required	Not Required	Not Required	120	40	400,000
	2,501–12,000 Not open to the public (Option 1)	Yes	Not Required	Not Required[e]	Not Required	120	40	400,000
	2,501–12,000 Not open to the public (Option 2)	Not Required[a]	Yes	Yes	Yes[h,i]	120	30[e]	200,000
	12,001–500,000	Yes	Not Required	Yes	Yes[h,i]	120	40	400,000
	Greater than 500,000[f]	Yes	Not Required	Yes	Yes[h,i]	120	40	400,000
High-hazard	0–500	Not Required[a]	Not Required	Not Required	Not Required	60	Not Required	Not Required
	501–2,500 Open to the public	Yes	Not Required	Not Required	Not Required	60	30	75,000
	501–2,500 Not open to the public (Option 1)	Yes	Not Required	Not Required	Not Required	60	30	75,000
	501–2,500 Not open to the public (Option 2)	Not Required[a]	Yes[g]	Yes	Yes[h,i]	60	20	50,000
	2,501–300,000	Yes	Not Required	Yes	Yes[h,i]	60	30	75,000
	300,001–500,000[f]	Yes	Not Required	Yes	Yes[h,i]	60	30	75,000

For SI: 1 foot = 304.8 mm, 1 cubic foot = 0.02832 m^3, 1 square foot = 0.0929 m^2.
a. Where automatic sprinklers are required for reasons other than those in Chapter 32, the portion of the sprinkler system protecting the high-piled storage area shall be designed and installed in accordance with Sections 3207 and 3208.
b. For aisles, see Section 3206.10.
c. Piles shall be separated by aisles complying with Section 3206.10.
d. For storage in excess of the height indicated, special fire protection shall be provided in accordance with note f where required by the fire code official. See also Chapters 51 and 57 for special limitations for aerosols and flammable and combustible liquids, respectively.
e. For storage exceeding 30 feet in height, Option 1 shall be used.
f. Special fire protection provisions, including but not limited to: fire protection of exposed steel columns; increased sprinkler density; additional in-rack sprinklers, without associated reductions in ceiling sprinkler density; or additional fire department hose connections shall be provided when required by the fire code official.
g. Not required where an automatic fire-extinguishing system is designed and installed to protect the high-piled storage area in accordance with Sections 3207 and 3208.
h. Not required where storage areas are protected by either early suppression fast response (ESFR) sprinkler systems or control mode special application sprinklers with a response time index of 50 (m • s)$^{1/2}$ or less that are listed to control a fire in the stored commodities with 12 or fewer sprinklers, installed in accordance with NFPA 13.
i. Not required in frozen food warehouses used solely for storage of Class I and II commodities where protected by an approved automatic sprinkler system.

For buildings containing high-piled combustible storage that are accessible to the public, such as mercantile (Group M) occupancies, the area thresholds for requiring automatic sprinkler protection are far less than the base values prescribed in IFC and Chapter 9 of the *International Building Code* (IBC). The requirement for the installation of an automatic sprinkler system is applicable in buildings open to the public storing Class I–IV commodities with a high-piled storage area over 2,500 square feet and high-piled storage areas over

FIGURE 14-12 If high-piled storage occurs in a Group M occupancy, automatic sprinkler protection is required when the storage area is greater than 2,500 square feet.

> **You Should Know**
>
> The occupancy classification does not affect the fire protection requirements for high-piled combustible storage. Fire protection requirements are based on
> - Commodity classification
> - Size of the designated high-piled combustible storage area
> - Height of storage ●

FIGURE 14-13 When a heat detection system is used for the protection of a high-piled storage area, it must be in accordance with NFPA 72.

500 square feet containing high-hazard commodities (Figure 14-12). By comparison, IFC and IBC Section 903.2.7 require automatic sprinkler protection when the fire area of a Group M occupancy exceeds 12,000 square feet. The intent of this provision is the protection of the building occupants. Group M occupancies can have large occupant loads and fuel packages. Fires involving high-piled combustible storage exhibit high heat release rates. Accordingly, when the Group M occupancy contains high-piled combustible storage, the IFC lowers the area threshold to ensure that design of the automatic sprinkler system can control or suppress an unwanted fire event. **[Ref. Table 3206.2]**

For smaller unsprinklered buildings that house high-piled combustible storage, a fire detection system is required when the high-piled combustible storage area contains Class I–IV commodities, has an area between 2,501–12,000 square feet and is not open to the public. Buildings not open to the public and containing high-hazard commodities require a fire detection system when the high-piled storage area is 501–2,500 square feet. Many buildings with this size of high-piled combustible storage area could be provided with an automatic sprinkler system for other reasons. Even though Table 3206.2 does not require the automatic sprinkler system, Footnote a states that it must be designed to protect the high-piled combustible storage. This concept is consistent with NFPA 13. In those buildings where an automatic sprinkler system is installed, Footnote g states that the fire detection system is not required. **[Ref. Table 3206.2]**

Where Group S occupancies lack climate control, design professionals usually will select heat detection as a means of complying with the fire detection requirement. NFPA 72, *National Fire Alarm and Signaling Code*, contains requirements for the installation of spot-type heat detectors in buildings with high ceilings. The requirements in NFPA 72 reduce the spacing of spot-type heat detectors as the height of the building increases (Figure 14-13). NFPA 72 also limits the use of these particular heat detectors to smooth, beam or sloped ceilings in a building 30 feet or less in height. Above 30 feet in height, NFPA 72 requires a performance design when spot-type heat detectors are specified unless a linear cable or pneumatic rate-of-rise heat detection system is used. Above 30 feet, other options could include the use of optical flame detection or linear beam smoke detection systems. **[Ref. 3206.5, Table 3206.2]**

For high-piled combustible storage, the IFC contains specific requirements for fire department access doors in buildings for manual fire-fighting and overhaul operations. Sprinklered buildings require access doors when the high-piled combustible storage area contains Class I–IV commodities over an area greater than 12,000 square feet or a high-hazard commodity over an area greater than 2,500 square feet. In buildings that are not sprinklered, the threshold for fire department access doors is 2,501 square feet for Class I–IV commodities and 501 square feet for high-hazard commodities, provided that the storage area is not open to the public. [Ref. Table 3206.2]

Fire department apparatus access roads are required to within 150 feet of all portions of the exterior walls of a building that can contain high-piled combustible storage. This is a similar requirement to Section 503.1.1 which is a general requirement for apparatus access roads, however, the exception to increase the distance when sprinklers are installed is not applicable to high-piled combustible storage. That exception is only found in Section 503.1.1. Section 102.10 states that when two code sections conflict with each other, the specific requirement supersedes the general requirement. In this case, the requirement in Chapter 32 is specific to high-piled combustible storage and would override the exception. [Ref. 3206.6]

Fire department access doors are required depending on the high-piled storage area and the commodities (Figure 14-14). Doors must be accessible without use of a ladder and spaced not more than 125 feet apart. The IFC also establishes requirements for minimum door dimensions, locking devices and limits on the type of door that can be used for access and prohibits the use of roll-up doors unless they are approved by the fire code official. [Ref. 3206.7]

FIGURE 14-14 Fire department access door in a building that contains high-piled combustible storage

One reason for the prohibition is that roll-up doors can be damaged by material-handling equipment, which could limit their use by fire fighters. Another concern is actually the task of forcing entry through a roll-up door. Generally, the task involves using a motorized saw, cutting an "X" into the door and pushing the cut metal into the building. Such tasks can lead to fire-fighter injuries. In 1999, data from the U.S. Fire Administration reported that the rate of fire-fighter injuries doubled when performing fire-fighting operations at commercial buildings, and 44 percent of all injuries were sprains/strains or bleeding/bruises/cuts or wounds.[9] It is for these reasons the fire code prohibits the use of roll-up doors for required access doors unless they are approved by the fire code official. [Ref. 3206.7.6]

---

[9] U.S. Fire Administration, Topical Fire Research Series, *Firefighter Injuries*, Volume 2, Issue 1, July 2001.

STORAGE METHODS

A number of storage methods are used in high-piled combustible storage, and these methods are regulated in IFC Sections 3207 through 3210. Each method of storage can present its own fire protection issues. The selected storage method is generally dictated by the stored commodity and economics. Certain storage methods may be limited in height based on the limitations of fire test data.

Solid-piled storage involves commodities that are moved without the use of material-handling aids, such as conventional or slave pallets (Figure 14-15). Solid-piled storage is commonly found for commodities such as rolled paper, rolled carpet or materials that are baled. Some lightweight commodities packaged in fiberboard cartons also may be stored using the solid-piled storage method. Materials stored in a solid pile are generally capable of supporting the dead load of additional materials added to the pile. Stored commodities are in direct contact with each other. Solid-piled storage is normally limited to storage on a floor of a building. Solid-piled storage has limited or no gaps or spaces that can serve as paths for air movement or fire spread. Fires involving solid-piled storage are the easiest to control in comparison to palletized or rack storage because they provide large, exposed surface areas to allow for the application of water and restrict air getting deeper into the pile.

FIGURE 14-15 Solid-piled storage of baled cotton

Pallets are designed to serve as a uniform platform for the packaging of commodities. They are constructed as a flat structure using wood, plastic, metal or paper. Palletized storage fires are more challenging to control in comparison to solid-piled storage (Figure 14-16). The pallet creates more surfaces available for ignition and allows for the horizontal spread of fire. By design, pallets have openings to accommodate the mechanical handling equipment. These openings create flue spaces that allow the horizontal spread of fire. Another consideration for fires involving palletized storage is the horizontal flue space created by the inherent hollow design of pallets is shielded from water discharged by an automatic sprinkler system. As a result, pre-wetting cannot occur because of the shielding. Because of these concerns with palletized storage, the IFC limits pile lengths for Class I–IV commodities to a maximum length of 120 feet and high-hazard commodities to a maximum pile length of 60 feet. [Ref. 3207.3, Table 3206.2]

FIGURE 14-16 Palletized storage

A variable that can influence the design of the automatic sprinkler system is plastic pallets used as a material-handling aid. Pallets, when constructed of combustible materials such as wood or plastic, represent a significant fire threat to a building and a very challenging automatic sprinkler system design. Pallet fires have very high heat release rates because of the large surface area to mass ratio. The IFC has specific requirements for plastic pallets stored inside of a high-piled combustible storage area. The IFC and NFPA 13 recognize pallets listed to UL 2335, Fire Tests of Storage Pallets, or FM 4996, Approval Standard for Classification of Pallets and Other Material Handling Products as Equivalent to Wood Pallets, as presenting a similar fire hazard as compared to wooden pallets (Figure 14-17). Pallets listed to this standard can be treated the same as wood pallets versus unlisted plastic pallets, which have the potential to change the hazard classification. Listed pallets can be identified by the listing mark that is molded into the pallet. Unlisted plastic pallets will increase the commodity class when they represent a more significant hazard than the product, see Table 14-2. [Ref. 3203.10]

FIGURE 14-17 This plastic pallet listed to UL 2335 is considered equivalent to a wood pallet with regard to fire load. *(Courtesy of Rehrig Pacific Co., Los Angeles, CA)*

TABLE 14-2 Effect of plastic pallet on commodity classification

Type of Plastic Pallet	Listed under UL 2335, or Approved under FM 4996	Nonlisted or Nonapproved Plastic Pallet
Unreinforced polyethylene or polypropylene	Treat as equivalent to wood pallet	Increase one commodity classification
Reinforced polyethylene or polypropylene	Treat as equivalent to wood pallet	For Class I, II or III – increase one commodity classification. For Class IV – treat as Group A cartoned, unexpanded plastic commodity
Plastic materials other than polyethylene or polypropylene	Treat as equivalent to wood pallet	Commodity classification based on specific fire testing, or increase two commodity classifications

Rack storage is the most predominant of all the methods of storage regulated by the IFC. Rack storage is commonly used because it can be erected rapidly, it is designed to facilitate a wide range of storage practices and its height is generally unlimited because of its inherent structural design as a metal frame. Storage racks are designed for mechanical or manual stocking and retrieval. A cursory review of NFPA 13 and Factory Mutual Global Loss Prevention Data Sheets finds that the greatest number of automatic sprinkler design options occurs in rack storage arrangements.

While it is not defined in the IFC or IBC, a storage rack is a combination of vertical, horizontal or diagonal structural members designed to support stored materials (Figure 14-18). Racks are constructed with or without solid shelves. Racks can be stationary, moveable or portable. Portable or moveable storage racks represent

FIGURE 14-18 Rack storage of commodities

FIGURE 14-19 Fire involving rack storage of motor oil *(Courtesy of International Code Consultants Inc., Austin, TX)*

their own fire protection engineering challenges and are outside the scope of this chapter. There are no provisions in the IFC or NFPA 13 that limit the height of storage racks. Storage racks can reach over 90 feet in height. It is a common construction practice to use the building storage rack as the load-bearing structural elements for the walls and roof when the building reaches these heights.

Of all of the storage configurations regulated in the fire code, rack storage of commodities represents the most challenging from a fire protection perspective. Fires involving rack storage (Figure 14-19) are one of the most challenging fires for an automatic sprinkler system to control or suppress because:

- The rack supports the commodity. This allows the commodity to be preheated and burn on the bottom, top and all vertical surfaces.
- A storage rack is constructed with airspaces on all sides that allows for rapid vertical and horizontal fire spread.
- The flue spaces between stored commodities allow for the velocity of the fire gases to be increased by compressing them between structural members and the exposed surfaces of the commodities. This increase of the gas velocity must be overcome by the momentum and mass of the water droplets from operating sprinklers.

Flue spaces are critical for the control or suppression of a fire involving commodities in fixed storage racks. Flue spaces are the open spaces between the commodities and the structural columns and beams of the rack—they are defined based on their orientation. The flue spaces that are perpendicular to the direction of loading the tiers with storage arrays are defined as longitudinal flue spaces. Longitudinal flue spaces traverse the entire length of the storage racks. Flue spaces between each rack upright column that parallel the direction of loading are transverse flue spaces (Figure 14-20). Flue spaces are essential in storage racks for high-piled combustible storage, because they permit ceiling sprinkler water to penetrate into commodities to either control or suppress the fire, depending on the design of the automatic sprinkler systems. **[Ref. 3208.3]**

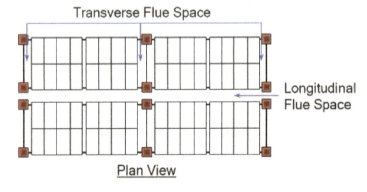

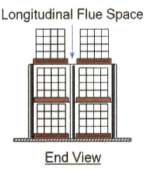

FIGURE 14-20 Longitudinal and transverse flue spaces serve to allow heat from the fire to rise to the sprinklers and allow water from the sprinklers to penetrate back down to the fire.

The IFC has requirements for automated storage. IFC Section 3209 contains requirements for automated storage applicable to carousel storage and automated rack storage. Automated rack storage is "a stocking method whereby the movement of pallets, products, apparatus or systems are automatically controlled by mechanical or electronic devices." [Ref. 202]

The method of automated storage can be either:
- Carousel storage systems, where all of the stored product moves on a support structure to the point where it is selected from the rack. The product can move in either a vertical or horizontal rotation (Figure 14-21); or
- Automated storage and retrieval systems, where individual products are moved into and out of a pallet storage rack. This typically involves computer controlled pallet movers and lifts (Figure 14-22).

FIGURE 14-21 Horizontal carousel storage system

Carousel storage systems are factory-built motorized storage systems that revolve around a fixed base. In most cases, the path of the revolution has two long parallel sides connected by round, short radius ends. They use fixed tracks with the motor mounted either on the top or bottom. A conventional carousel storage system revolves in the horizontal plane, but may be vertical, or a combination of both. Products stored in the carousel are brought to a stationary picking station using manual or computer control. Computers are commonly used to simultaneously maintain inventory records and conduct ordering. Typically, the commodities are small parts and products. [Ref. 3209.3]

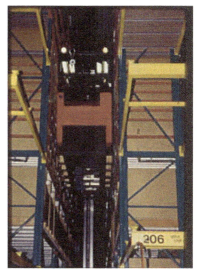

FIGURE 14-22 Automated rack storage system

Automated storage and retrieval systems, commonly referred to as ASRS, consist of a rack storage arrangement with computer-controlled pallet movers and product lifts. The height of an automatic storage and retrieval system in rack storage can often reach over 100 feet. Typically, the aisles are only wide enough for the movement of a pallet, and the automated lift can move both horizontally from one end of the aisle to the other and vertically to the upper storage tiers. These systems require shutdown of the pallet lifts during a fire to prevent additional product being moved into the fire location, burning product being moved to a location not yet involved in the fire, and to protect the fire fighters as they move down the aisles. [Ref. 3209.4]

For horizontal carousels and ASRS, NFPA 13 may require the installation of an in-rack automatic sprinkler system, especially if the height of the storage is over 8 feet and combustible containers are used. [Ref. 3209.2]

Because carousels rotate on a fixed track, a significant concern is a fire involving the carousel system moving the fire throughout an area. To limit the potential for such an incident, automatic shutdown features are prescribed. These include either the installation of an automatic smoke detection system or a dead-man switch. If the smoke detection system is activated or the dead-man switch is released, the carousel ceases to operate and allows the sprinkler system to activate and attack the fire. [Ref. 3209.3]

AISLES

The requirements for aisles are applicable when the high-piled combustible storage area is greater than 500 square feet. Aisles are provided to facilitate material handling and provide access for firefighting and overhaul activities. The requirements for the minimum width of the aisles are based on whether the building is or is not protected by an automatic sprinkler system. Aisles are required to comply with the requirements of NFPA 13 based on the method of storage and the design of the automatic sprinkler system. When applying the requirements of NFPA 13, an important variable is the minimum aisle width between storage piles or racks. Aisles are required by IFC Chapter 10 to facilitate means of egress from a building. [Ref. 3206.10]

A primary role of aisles is to limit the spread of fire via radiant heat transfer. Aisles reduce the potential for fire spread from one storage pile or rack to another by providing a defined space between these areas. Radiant heat transfer is the result of electromagnetic radiant energy that arises due to the temperature of a source. Because it is an electromagnetic source of radiation, heat transfer occurs without the need for air or metal. A classic example of radiant heat transfer is sunlight. When a cloud covers the sun, both its heat and light diminish. Aisles provide a mechanism to reduce the potential for a fire to transfer enough energy to ignite adjacent storage piles or racks. At a minimum, NFPA 13 requires rack storage to have 4-foot aisles. As the width of aisles increases, the required sprinkler discharge is reduced because the potential for radiant heat transfer is further reduced. A basic rule of radiation heat transfer is that for a reduction of one-half in aisle width, the amount of radiant heat transfer is doubled when evaluating a fire in a rack or storage pile and its ability to spread to an adjacent rack by crossing over the aisle. And conversely, when the width of the aisle is doubled, the radiant heat is reduced by half. This rule of radiation heat transfer serves as an important reason for verifying the width of aisles, especially when inspecting buildings housing high-piled combustible storage (Figure 14-23).

Buildings that are not protected by an automatic sprinkler system must have a minimum width of 8-foot (or 96-inch) aisles. Buildings that are equipped with sprinklers may be allowed to have aisle widths of less than 8 feet. However, this is dependent on the design of the automatic sprinkler system, the stored commodity, the storage method and if the building is accessible to the public. The IFC allows aisles limited to employee access to be as little as 24 inches in width. [Ref. 3206.10.1]

You Should Know

High-piled combustible storage can create a serious fire and life safety hazard. With proper design, installation and maintenance, the hazards created by high-piled combustible storage can be adequately addressed to provide a safe operation for the employees, the public and the responding fire fighters. •

FIGURE 14-23 Aisles allow for movement of material-handling equipment, provide an exit access path, limit the potential for fire spread between racks and provide fire-fighting access.

Removing or inserting products on racks (stocking) may be done manually or mechanically. Manual stocking employs ladders or other nonmechanical equipment to move stock, while mechanical stocking uses motorized vehicles such as fork lift trucks and pallet trucks. Requirements are specified for the maintenance of aisles that stipulate the minimum aisle width required during periods when manual or mechanical commodity stocking is underway. Aisles, exit doors and fire department access doors cannot be obstructed and shall be maintained free of obstructions (Figure 14-24). For manual stocking operations, aisle widths must equal at least 50 percent of the aisle dimension for aisles wider than 48 inches. For aisles 48 inches or less in width, a minimum 24-inch aisle is required during manual stocking operations. In cases where mechanical stocking occurs, a minimum 44-inch aisle is required. [Ref. 3205.4]

FIGURE 14-24 Aisles must be maintained during stocking operations and cannot obstruct the exit access.

CHAPTER 15
Other Special Uses and Processes

The *International Fire Code* (IFC) regulates numerous hazardous and special processes and uses. Several of these activities present a potential for fires and explosions if the use or process is improperly designed, operated or maintained. In the case of the construction and demolition of a building, these activities can compromise fire-fighter safety because structural components and fire protection systems are either incomplete or compromised. Activities involving the use of combustible dusts present the hazard of a deflagration that can injure and kill plant personnel. Hot work is a hazardous activity because it is a source of ignition by the nature of the activity. Marijuana grow and extraction facilities use carbon dioxide and flammable gases which present significant hazards to employees and fire fighters. The special processes and uses reviewed in this chapter are combustible dust producing operations, fire safety during construction and demolition, hot work and grow and processing facilities for marijuana.

COMBUSTIBLE DUST-PRODUCING OPERATIONS

Combustible dust-producing operations occur in a variety of industries, including food production, manufacturing of pharmaceuticals, certain woodworking operations and some plastic manufacturing processes. IFC Chapter 22 addresses the prevention of dust explosions, which technically are dust deflagrations. A deflagration is an exothermic reaction (meaning it releases heat) resulting from a rapid oxidation of a combustible dust, in which the reaction progresses through the unburned material at a rate less than the velocity of sound. Deflagrations are commonly termed as "slow explosions." Deflagrations are far more common than explosions, which have burning rates greater than the speed of sound. [Ref. 202]

A dust deflagration requires a combustible dust as the source of fuel. A combustible dust is a "finely divided solid material which is 420 microns or less in diameter, which when dispersed in air in the proper proportions, it can be ignited by a source of ignition. Combustible dust will pass through a U.S. number 40 sieve." Particle size is important. When particles become smaller, their mass is reduced, causing their surface area to become proportionally larger as compared to their mass. This increases the potential energy in a dust deflagration because the material is more easily ignited with combustion and total consumption of the fuel particle occurring nearly instantaneously. [Ref. 202]

A dust is combustible when its particles will burn. Table salt is small enough to be considered a dust, but it is chemically noncombustible. Wood is combustible, but dimensioned lumber is not dust: it is a solid mass with a surface area large enough to make it difficult to ignite in air using an oxygen-acetylene torch. In order for the provisions in IFC Chapter 22 to be applicable, the material must meet the definition of a dust (particle size), and the material must be combustible. At a cabinet shop, one would expect to find combustible dust from that dimensional lumber. Not all of the waste is dust—table saw waste for example would be referred to as "chips" in the IFC since it is larger than 420 microns. Dust would be produced in sanding operations and possibly planning operations.

The energy required to ignite a combustible dust is defined as the minimum ignition energy and is measured in millijoules. A typical spark created by walking across a carpeted floor and touching a metal door is about 100 millijoules. The lower the minimum ignition energy value, the less energy is required for ignition to occur. Consider, for example, an agricultural dust such as wheat flour with an average particle size of 80 microns. The minimum ignition energy required to ignite such dust is approximately 95 millijoules. If the particle size is doubled to 160 microns, the required ignition energy is over 400 millijoules. Table 15-1 lists the particle size of common materials.

> **You Should Know**
>
> Even though the process of a deflagration is slower than the speed of sound, and slower than a detonation, when the deflagration occurs inside a building, container or confined space, the results can be equally devastating. •

TABLE 15-1 Particle sizes of common materials

Material	Size (micron)
White granulated sugar	450–600
Sand	50 and greater
Talcum powder	10
Mold spores	10–30
Human hair	40–300
Wheat, corn or soybean flour	1–100

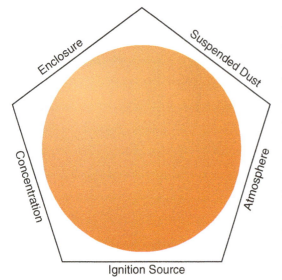

FIGURE 15-1 Dust deflagration pentagon

FIGURE 15-2 Dust collector used to capture combustible dust produced during the manufacturing of furniture

The mechanism of a dust deflagration requires more conditions when compared to the conventional fire triangle. Along with an ignition source and proper mixing of the fuel with an oxidizer, a dust deflagration also requires the fuel be in a confined enclosure, such as a building or exhaust duct, and it must be easily dispersed in the atmosphere within the enclosure. A deflagration can occur when enough dust particles are suspended in the enclosure, the concentration exceeds the minimum explosive concentration (MEC) and the ignition source is greater than the minimum ignition energy (MIE). The burning combustible dust liberates flammable gases and it is the ignition of these gases that causes the deflagration (Figure 15-1).

The IFC specifies requirements for controlling sources of ignition and housekeeping to reduce the potential of a dust deflagration. The IFC adopts NFPA standards that regulate dust deflagration hazards. To prevent the ignition of the dust layer, the IFC requires strict control of ignition sources. A means of collecting the dust is required to prevent the development of a dust accumulation that can be suspended in air. The key is to ensure that the dust does not accumulate at any locations where it can be suspended in air, such as on ventilation ducts, light fixtures, building trusses or purlins, cable trays or similar locations (Figure 15-2). **[Ref. 2203.1, 2203.2]**

In two of the referenced NFPA standards adopted by the IFC, one means of determining if a dust deflagration hazard exists is to measure the dust layer depth. NFPA 654, Standard for the Prevention of Fire and Dust Explosions from the Manufacturing, Processing and Handling of Combustible Particulate Solids, and NFPA 664, Standard for the Prevention of Fires and Explosions in Wood Processing and Woodworking Facilities, indicate that such a hazard exists when the dust is combustible, it has a density of 75 pounds/cubic foot or less, and the area of dust layer and its depth exceeds the values indicated in Table 15-2.

TABLE 15-2 Minimum dust layer depth and area for a dust deflagration hazard by material or facility type

Material or facility type	Dust layer depth	Dust layer area
Combustible particulate solids	$1/32$ inch	• For buildings less than 20,000 square feet, the area of the dust layer shall not exceed 5% of the floor area • For buildings of 20,000 square feet or greater, the area of the dust layer shall not exceed 1,000 square feet
Wood processing and woodworking facilities	$1/8$ inch	

There are many materials that in a fine form can create a dust deflagration; i.e. coal, wood, sulfur, grain, pharmaceuticals and combustible metals. Each of the materials has different specific mitigation measures. The IFC addresses these various hazards by requiring a dust hazard analysis as required by NFPA 652, Standard on the Fundamentals of Combustible Dust. [Ref. 2203.1]

If, based on the dust hazard analysis, a dust deflagration potential exists, then specific NFPA standards must be complied with to mitigate the hazards. These various standards are shown in Table 15-3. Typical areas of concern in these facilities are routine housekeeping and adequate dust collection. Fugitive dust will settle on horizontal ledges and becomes airborne when dislodged; if the concentration is above the MEC and a source of ignition is available, a dust deflagration will occur (Figure 15-3).

FIGURE 15-3 Agricultural dust has settled on every horizontal surface available. Based on Table 15-2, since this dust is more than $1/32$ inch deep, if it covers more than 5 percent of the floor area there is adequate fuel for a dust deflagration.

TABLE 15-3 Industry specific standards

Standard	Subject
NFPA 61	Standard for the Prevention of Fires and Dust Explosions in Agricultural and Food Processing Facilities
NFPA 69	Standard on Explosion Prevention Systems
NFPA 70	*National Electric Code*
NFPA 85	*Boiler and Combustion System Hazards Code*
NFPA 120	Standard for Fire Prevention and Control in Coal Mines
NFPA 484	Standard for Combustible Metals
NFPA 654	Standard for Prevention of Fire and Dust Explosions from the Manufacturing, Processing and Handling of Combustible Particulate Solids
NFPA 655	Standard for the Prevention of Sulfur Fires and Explosions
NFPA 664	Standard for the Prevention of Fires and Dust Explosions in Wood and Woodworking Facilities

FIRE SAFETY DURING CONSTRUCTION AND DEMOLITION

Erecting or demolishing a building introduces a variety of hazards. When constructing a building, unprotected shafts may be created that form a vertical path for fire travel. A variety of hazardous materials can be on site such as paints, waste materials and fuels that can contribute to fire growth and spread. When a building is demolished, structural components may be weakened or removed, which reduces the building's resistance to external loads like rain, wind, snow or ground motion from an earthquake. Previously functioning fire protection systems (sprinklers and standpipes) may be disconnected or otherwise inoperable. NFPA reports that fire departments in the U.S. annually respond to an estimated 3,750 fires in buildings under construction.[10]

There have been several significant fires in buildings under construction within the last decade, some of which are listed in Table 15-4. IFC Chapter 33 addresses the fire safety aspects of structures being constructed, altered or demolished. These requirements address temporary heating, establishing precautions against fire, the storage and use of hazardous materials and maintaining fire protection systems. [Ref. 3301.1]

TABLE 15-4 Significant fires during construction

Date	Location	Description
December 2014	Los Angeles, CA	7-story, 526 units, residential structure
August 2013	Portland, OR	5-story, 46 units, residential structure
July 2017	Oakland, CA	7-story, 196 units, residential and retail structure
March 2017	Raleigh, NC	5-story, 241 units, residential structure
February 2017	Maplewood, NJ	4-story, 200+ units, residential structure
March 2017	Overland Park, KS	6-story, residential structure, 18 other structures damaged
June 2017	East Hollywood, CA	3-story, residential structure

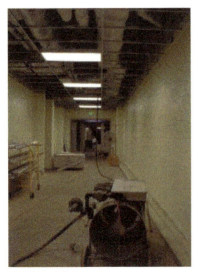

FIGURE 15-4 Heaters must be used and maintained based on the requirements in the IMC and IFGC.

Temporary heating is commonly employed in buildings during construction or demolition to maintain water-based fire protection systems above water's freezing temperature and the safety of personnel. Portable heaters used to heat the building are required to be listed, and installed with the *International Mechanical Code* (IMC) and the *International Fuel Gas Code* (IFGC). Depending on the fuel, refueling must be in accordance with applicable IFC requirements. Clearances between the heater and combustibles must be maintained to avoid ignition of construction materials. Heaters must be supervised by competent personnel (Figure 15-4). [Ref. 3303]

---

10. Campbell, Richard, "Fires in Structures Under Construction, Undergoing Major Renovation, or Being Demolished," April 2017, p.iii, NFPA, Quincy, MA,

Another consideration during building construction or demolition is the type and amount of fuels available on-site for equipment and machinery. The predominance of fueled equipment found at these sites results in storage of No. 2 diesel fuel, unleaded gasoline or LP-gas. The requirements in IFC Chapter 33 require compliance with the IFC flammable and combustible liquid and LP-gas provisions. Any operations that require the application of flammable and combustible liquids, such as the application of paints, varnishes and coatings, require these activities be performed in areas with adequate ventilation (Figure 15-5). Note that specific requirements for such activities are also found in IFC Chapter 24 for flammable finishing. [Ref. 3305]

FIGURE 15-5 Interior application of flammable finishes in buildings under construction or renovation is a regulated activity.

A major consideration during building construction, alteration and demolition is ensuring the contractor and subcontractors are properly trained, aware of the fire hazards found on a construction site, and know what actions are required in the event that an unwanted fire occurs. The responsibility for implementing a fire prevention plan lies with the owner or any person the owner designates as the fire prevention program superintendent. Often this responsibility falls to the project superintendent. The fire prevention program superintendent is responsible for the maintenance of any required fire protection systems and is the impairment coordinator. The superintendent is responsible for training job-site personnel about the building's fire protection features and systems and how they are maintained and serviced. An approved prefire plan must be prepared for fire fighters to indicate the location and types of various hazards or pitfalls the building may present. The fire prevention program superintendent is responsible for the implementation of the hot work permit program (see the next section). [Ref. 3308]

The prefire plan should address the installation and maintenance of fire department access, which includes vehicle access to the site and building, particularly for a new development, and access into the building for both new construction and alterations. Access to the building and site can be accomplished with the installation of a key box at a temporary gate or entrance. Often, temporary roads or driveways may be used during the construction process. Even temporary access is required to meet the approval of the fire code official and must be maintained clear and unobstructed. Whether temporary or permanent, the access roads must have road signs to designate street names for emergency responders. Water supply for fire protection must be provided during construction and demolition. Section 507 would be used to determine the required water supply for new construction. Section 3312.1 states that either the permanent water supply can be installed prior to combustible materials being brought on site, or a temporary water supply can be provided where approved. Frequently, since the permanent water supply is part of the construc-

Code Essentials

Owner's responsibility
- Develop and implement a fire prevention program
- Designate a fire prevention program superintendent (if none designated, then it is the owner by default)
- Develop a prefire plan in cooperation with the fire chief
- Ensure fire protection equipment is functional and available
- Supervise the hot work program
- Train personnel
- Maintain training records

tion project, a temporary water supply is approved by the fire code official so construction can commence until the permanent water supply is completed, tested and approved. [Ref. 505, 506, 3310, 3312]

Maintaining fire protection systems and features is an important element during construction, alteration or demolition of a building. Buildings under construction or demolition involved in a fire have the greatest potential for collapse or other failures because the fire and life safety systems are not completed or are being demolished. The IFC requires standpipe systems complying with NFPA 14, Standard for the Installation of Standpipe and Hose Systems, be provided when construction height is over 40 feet. The standpipe system must be extended to within one floor of the highest level of construction where secured flooring or decking is present. During demolition, the standpipe system is required to be maintained within one floor of the floor level being demolished. Similar requirements exist for automatic sprinkler systems (Figure 15-6). In addition, the IFC specifically states that in buildings where an automatic fire sprinkler system is required, the system must be complete, tested and approved prior to any occupancy of the building being allowed. [Ref. 3313, 3314]

FIGURE 15-6 Combination standpipe and sprinkler fire department connection

The fire code official is authorized to require a fire watch during construction. Likewise, the prefire plan developed with the fire chief may also make the requirement for fire watch. The IFC provides some guidance on the fire watch duties by stating that it shall occur during nonworking hours for new construction that exceeds 40 feet in height. Fire watch personnel must be trained, and fire watch is their only duty, although it is acceptable to provide fire watch and security. Fire watch personnel must make constant patrols and have the ability to notify the fire department. [Ref. 3304.5]

WELDING AND OTHER HOT WORK

The construction of buildings, machinery, appliances and industrial processes commonly requires the use of welding or other hot work processes to coalesce various ferrous and nonferrous metals (Figure 15-7). Hot work using oxygen and a fuel gas is performed to cut these metals into shapes or objects. All of these and other activities, including the installation of torch-applied roofing systems or thawing of frozen water pipes, constitute hot work and are regulated by the IFC. [Ref. 202]

FIGURE 15-7 Hot work such as welding creates a fire hazard during construction, demolition, maintenance and repair.

The IFC regulates hot work activities based on the type of operation performed, such as electric arc welding, gas welding and cutting or a torch-applied roof system. These operations must satisfy the general and fire safety requirements in IFC Chapter 35. Under the IFC, an operational hot work permit is issued when the permit applicant presents an acceptable plan for managing this activity. The permitted party is then responsible for all hot work operations within their facility or building for the life of the operating permit. For certain activities such as hot work causing the impairment of an automatic sprinkler system, the potential exists for the ignition of flammable vapors or combustible materials or on board of ships at dock, specific approval by the fire code official is required. [Ref. 3501.3]

All hot work operations must be performed under the supervision of a person responsible for the hot work program. The individual is responsible for reviewing the site prior to issuing a permit and performing subsequent inspections as work progresses to ensure it is in accordance with the hot work permit program. This includes maintaining adequate documentation demonstrating compliance with IFC Chapter 35, the hot work plan and ensuring that adequate signs are posted in the work area (Figure 15-8). [Ref. 3503.3]

FIGURE 15-8 Sign prohibiting hot work in a flammable vapor area

An area review and inspection is required prior to commencing hot work. This inspection, performed by the responsible person, ensures combustibles and building openings are properly shielded and that the area is clean of any combustible debris. When welding partitions are used, the inspection should verify the partitions prevent the passage of sparks, slag and heat from the hot work area. In areas protected by an automatic sprinkler system, the sprinklers in the immediate area can be shielded; however, these shields must be removed at the end of the work day or assignment. [Ref. 3504.1, 3504.3]

A fire watch is required during and after hot work activities, and is maintained for at least 30 minutes after hot work has concluded (Figure 15-9). The individual is responsible for extinguishing any spot fires and communicating an alarm to the fire department. Fire watch personnel must be trained in the use of portable fire extinguishers. A fire watch is not required in areas with no fire hazards or no combustible materials. [Ref. 3504.2]

FIGURE 15-9 Fire watch during hot work operations

FIGURE 15-10 Metal cutting using oxygen-acetylene fuel gas

FIGURE 15-11 Oxygen-acetylene fuel gas cutting torch

The use of oxygen along with a flammable gas is common in many cutting, welding and brazing activities. Typical fuel gases include acetylene, methylacetylene-propadiene (MAPP gas) or LP-gas (Figure 15-10). Of these three gases, acetylene is the most commonly used because of the high temperatures it can develop when it is properly mixed with oxygen. Oxygen-acetylene fuel gas mixtures can produce temperatures from 5,800 to 6,300°F depending on the mixture. These high temperatures can cut through a variety of metals. Acetylene is classified as compressed flammable and Class 2 unstable (reactive) gas. The gas is dissolved in solution with acetone to facilitate its safe storage and handling.

An oxygen-acetylene torch is assembled using two compressed gas cylinders (Figure 15-11). One cylinder contains compressed or cryogenic oxygen and the other cylinder contains fuel gas. Because the gases are stored at high pressures, a pressure regulator is installed on each cylinder to control the pressure at the torch. Downstream of the regulator discharge fitting, a hose is connected that is terminated at the cutting torch. While oxygen and all of the fuel gases are incompatible hazardous materials (see Chapter 16 of this text), the IFC allows the cylinders to be located adjacent to each other because they are equipped with pressure regulators and have a low loss history involving oxygen-fuel gas cutting and welding. **[Ref. 3505.2.1]**

The appropriate pressure regulator is required for oxygen and the selected fuel gas. Acetylene pressure regulators are designed to limit the discharge pressure to 15 PSIG or less (Figure 15-12). Acetylene becomes explosively reactive above this pressure so the IFC establishes a 15 PSIG pressure limit. Oxygen regulators must be designed and handled so they never come into contact with any flammable or combustible liquids, including lubricants such as oil. Oil and oxygen are incompatible hazardous materials, and if oil is ever introduced into an oxygen system, the pressure regulator can explode. **[Ref. 3505.3, 3505.4]**

FIGURE 15-12 Acetylene pressure regulator *(Courtesy of Smith Equipment, Watertown, SD)*

Another common hot work activity is the installation of torch applied roofing systems (Figure 15-13). The roofing materials are made water tight by fusing and melting bitumen and asphalt in the roofing material to the roof's substrate. The fusing of these materials is commonly performed using specialized LP-gas fueled burners. Torch-applied roofing systems should only be performed by individuals who have been trained in the fire safety requirements for the application of these materials, the safe storage and handling of LP-gas, and in the use of portable fire extinguishers. [Ref. 3317.1]

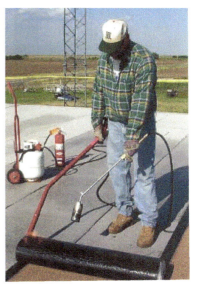

FIGURE 15-13 Installation of a torch-applied roofing system *(Courtesy of Midwest Roofing Contractors Association, Lawrence, KS)*

HIGHER EDUCATION LABORATORIES

Higher education institutions are quite possibly one of the most important industries to the advancement of technology, science and medicine. These academic institutions often have chemistry, biology, medical, engineering and other laboratories where hazardous materials are used. These laboratories contain hazardous materials, and the code official tries to apply the general hazardous materials provisions (more details of the general hazardous materials provisions can be found in Chapter 16 of this text) which oftentimes are not appropriate for specialized academic laboratory settings.

Chapter 38 is designed to apply to these facilities and takes into account the unique circumstances found in these laboratories, such as

- Lower chemical density
- Staff oversight
- Mixed-use occupancies.

The provisions in Chapter 38 apply to higher education laboratories, defined as "laboratories in Group B occupancies used for educational purposes above the 12th grade. Storage, use and handling of chemicals in such laboratories shall be limited to purposes related to testing, analysis, teaching, research or developmental activities on a nonproduction basis." [Ref. 202]

Higher education laboratories have the option of designing using the control area concept (see "control areas" in Chapter 16 of this text), using the laboratory suite concept, or classification as a Group H occupancy (Figure 15-14). The Group H classification is difficult for multistory buildings since the height of Group H is limited in all but Type IA construction. The laboratory suite concept allows labs to exist and operate in high-rise buildings, but it is limited to higher education facilities. Which means if a lab is located in a Group B office building, the concept of laboratory suites cannot be applied.

FIGURE 15-14 Laboratory at a higher education campus

> **Code Essentials**
> - Laboratory suites are a design alternative to control areas. Control areas can be used in higher education laboratories, but the allowable quantities are different.

The laboratory suite concept allows more flexibility in hazardous materials use and increased quantities, while adding administrative and construction safeguards. Laboratory suites are separated from each other with fire-resistance-rated construction, and are also required to be separated from the non-laboratory areas of the building. The number of laboratory suites per floor is limited, and the quantity of hazardous materials is limited. Table 15-5 shows the allowable laboratory suites and the allowable percentage of the maximum allowable quantities. These percentages are greater than those found in Section 5003.8.3.2, but are only allowed when the additional safeguards are provided.

TABLE 15-5 Design and number of laboratory suites per floor (IFC Table 3804.1.1)

Floor level		Percentage of the maximum allowable quantity per lab suite[a]	Number of lab suites per floor	Fire-resistance rating for fire barriers in hours[b]
Above grade plane	Higher than 20	Not allowed	Not allowed	Not allowed
	16–20	25	1	2[c]
	11–15	50	1	2[c]
	7–10	50	2	2[c]
	4–6	75	4	1
	3	100	4	1
	1–2	100	6	1
Below grade plane	1	75	4	1
	2	50	2	1
	Lower than 2	Not allowed	Not allowed	Not allowed

a. Percentages shall be of the maximum allowable quantity per control area shown in Tables 5003.1.1(1) and 5003.1.1(2), with all increases allowed in the footnotes to those tables.
b. Fire barriers shall include walls, floors and ceilings necessary to provide separation from other portions of the building.
c. Vertical fire barriers separating laboratory suites from other spaces on the same floor are permitted to be 1-hour rated.

The quantities in Table 15-5 (IFC Table 3804.1.1) provide a percentage of the quantities in Table 5003.1.1(1) and 5003.1.1(2). See Table 16-3 in this text, page 234, for Table 5003.1.1(1). Footnote a on Table 3804.1.1 states that all of the increases allowed by Table 5003.1.1(1) can also be applied. For example, consider a sprinklered, 7-story building at a university with laboratory suites on the sixth floor. Four laboratory suites are allowed on the sixth floor, separated by 1-hour fire-resistance-rated fire barriers. Each lab is using Class IB flammable liquids, each laboratory suite can contain 75 percent of the maximum allowable quantity. Table 5003.1.1(1) specifies that the maximum allowable quantity in storage is 120 gallons, which can be doubled to 240 gallons with sprinklers. For the laboratory suite, only 75 percent is allowed; therefore, the quantity in each lab is limited to 180 gallons. **[Ref. Table 3804.1.1]**

There is also a density limitation for Class I flammable liquids, which is
- 8 gallons per 100 square feet on floors 1 through 3,
- 6 gallons per 100 square feet on floors 4 through 6 and
- 4 gallons per 100 square feet on floors 7 and above.

The previous example determined that 80 gallons of Class IB is allowed in each laboratory suite on the sixth floor, but to actually have that quantity, each laboratory suite must be at least 4,500 square feet. Hazardous material containers are limited in size to a maximum of 5.3 gallons, 50 pounds or 100 cubic feet. So, the 180 gallons must be stored in containers not exceeding 5.3 gallons each. [Ref. 3803.2.1, 3803.2.2]

EXERCISE

A university campus has laboratory suites on the eighth floor of a sprinklered, Group B building of Type IB construction. Each lab desires to have a solid Class 2 Oxidizer.

Calculate the following:
1. The number of laboratory suites allowed on the 8th floor.
2. The minimum required fire separation between each laboratory suite.
3. The maximum allowed quantity in each laboratory suite.

SOLUTION

1. The number of laboratory suites are allowed on the 8th floor.
   a. Table 3804.1.1 indicates that two laboratory suites are allowed on the 8th floor.
2. The minimum required separation between each laboratory suite.
   a. Table 3804.1.1 indicates that the laboratory suites must be separated by 2-hour fire barriers.
3. The maximum allowed quantity in each laboratory suite.
   a. Table 3804.1.1 indicates that 50 percent of the maximum allowable quantity per control area is allowed.
   b. Table 5003.1.1(1) indicates that the maximum allowable quantity per control area is 250 pounds.
   c. Footnote d allows a 100 percent increase where sprinklered, so 250 × 2 = 500 pounds.
   d. 500 × 0.50 = 250 pounds

The examples above are based on laboratory suites that are in sprinklered buildings. Note that where the laboratory suites exist in nonsprinklered buildings, a different set of criteria is applied. Without the automatic sprinkler system, the quantities are reduced, the number of laboratory suites are reduced and the fire-resistance-rated separations are increased. Additionally, an automatic fire detection system shall be installed. [Ref. 3805.2]

The requirement for a fire detection system is a retroactive requirement. Even though it is not found in Chapter 11 for retroactive construction requirements, it applies to all existing laboratory suites in existing nonsprinklered buildings. [Ref. 3805.3]

Whenever a new laboratory suite is constructed it must be located in a sprinklered building. If an existing, nonsprinklered building is adding, or creating, a new laboratory suite, the building must be equipped with an automatic sprinkler system. [Ref. 3803.1.7]

Additional safeguards are added to allow these increased quantities. Standby and emergency power is required for laboratory suites located above the sixth floor or below grade. A chemical safety review is required which addresses all operating and emergency procedures for the laboratory suite. [Ref. 3803.1.1, 3804.1.1.6]

MARIJUANA GROW AND PROCESSING FACILITIES

Many jurisdictions are considering, or have already passed legislation, legalizing the production, processing and sale of marijuana. Aside from the drug implications, there are several fire and life safety issues that are raised in all of these processes, including the use and heating of flammable liquids or gases, heat and electrical hazards and the practice of using carbon dioxide to enhance growth. The IFC regulates these facilities and operations, although the code does not refer to marijuana. The code provides requirements for plant processing and extraction facilities in Chapter 39 and carbon dioxide enrichment in Section 5307.4. By not referring specifically to marijuana and cannabis oil, these regulations apply to marijuana and also to any other operation utilizing the same processes because the hazard will be the same.

Extraction processes typically use flammable or combustible liquids or gases to extract the oils from the plant. The flammables act as a solvent to capture the oils. Then the liquid is heated and distilled to separate the oils from the extraction solvent. In the case of marijuana, cannabis oil is rendered and captured. These processes require a construction permit and an operational permit (Figure 15-15). [Ref. 105.6.38, 105.7.18]

FIGURE 15-15 Extraction process utilizing LP-gas as the solvent

Because of the hazards in the extraction process, it must be located within a room dedicated to extraction and cannot be located in a building containing Groups A, E, I or R occupancies. The extraction equipment must be listed, or a technical report must be provided. The technical report must address the specific equipment and vessels and provide a process hazard analysis. Once the equipment is installed and prior to inspection by the fire code official, the engineer of record or other approved professional shall inspect the system to ensure the equipment and design matches the technical report. [Ref. 3903.2, 3904]

A gas detection system is required where the extraction process utilizes flammable gas as the solvent. An emergency shutoff is required for extraction processes utilizing any gaseous hydrocarbon-based solvent. [Ref. 3905.1, 3905.2]

Carbon dioxide can be used to stimulate plant growth. Carbon dioxide enrichment is a process where carbon dioxide gas is intentionally introduced into an indoor environment, typically for the purpose of stimulating plant growth. All plants need carbon dioxide to grow and be healthy, by adding carbon dioxide into a room the rate of plant growth can be enhanced. However, these systems also present concern because these rooms are occupied rooms. Adding carbon dioxide into an occupied room must be monitored carefully. Carbon dioxide exists in the air we breathe at a level around 400 parts per million (ppm). The typical level of carbon dioxide in a grow facility to stimulate plants is in a range of 1,000 – 1,500 ppm. Additional carbon dioxide is added to the grow room when the lights are on—the daylight growing period. This level must be monitored so it is maintained below the PEL. [Ref. 5307.4.3]

A carbon dioxide enrichment system is required to obtain a permit when the quantity exceeds 875 cubic feet (100 pounds). An operational permit is required for storage of compressed carbon dioxide for enrichment systems at that quantity. The requirements in Section 5307.4 apply to any carbon dioxide enrichment system over that quantity, or to any carbon dioxide enrichment system with a remote fill connection (Figure 15-16). [Ref. 105.6.8, 5307.4]

The installation of carbon dioxide enrichment systems must meet the requirements in Chapter 33 and NFPA 55, *Compressed Gases and Cryogenic Fluids Code*. A gas detection system is required for these systems. The gas detection system is to alarm at two levels; low-level alarm at 5,000 ppm (0.5 percent), and a high-level alarm at 30,000 ppm (3 percent). The permissible exposure limit established by OSHA for carbon dioxide is 5000 ppm (0.5 percent concentration). 30,000 ppm represents the Threshold Limit Value – Short Term Exposure Limit (TLV-STEL). Both alarm thresholds result in stopping the flow of carbon dioxide and activating the mechanical exhaust system, but the low-level threshold activates a supervisory signal at an approved location, while the high-level threshold activates an audible and visible alarm inside and outside of the carbon dioxide enrichment area. The exhaust system must provide a minimum of 1 cfm per square foot with exhaust intakes within 12 inches of the floor. [Ref. 5307.4.3, 5307.4.4]

FIGURE 15-16 Carbon dioxide enrichment systems require a permit where the quantity of carbon dioxide is over 100 pounds.

You Should Know

The IFC contains 19 chapters that address the hazards in specific uses and occupancies. These requirements apply, in most cases, regardless of the occupancy classification because they deal with the operation or use occurring in the building.

PART VI
Hazardous Materials

Chapter 16: General Requirements for Hazardous Materials

Chapter 17: Compressed Gases

Chapter 18: Flammable and Combustible Liquids

CHAPTER 16
General Requirements For Hazardous Materials

Requirements for storage, handling, use and dispensing of hazardous materials represent the largest body of regulations in the *International Fire Code* (IFC). Hazardous materials can be a challenge because their proper classification is essential in determining and applying the IFC and *International Building Code* (IBC) requirements. In some cases, determining the proper classification itself can be a significant challenge that requires technical assistance from a qualified person. Not all chemicals and chemical compounds are hazardous materials as defined by the IFC. Chemicals and chemical formulations that do not fall into one of the 12 IFC hazard categories are exempt from regulation as a hazardous material under the IFC and IBC. Currently, the American Chemistry Society has over 133 million chemicals or formulations registered in the Chemical Abstracts Service. The IFC regulates only a small percentage of these chemicals; however, the chemicals that are regulated are essential in the manufacturing of numerous consumer and industrial products.

Chapter 16 reviews the general requirements for hazardous materials. IFC Chapter 50, Hazardous Materials—General Provisions, is applied in conjunction with the material-specific requirements in IFC Chapters 51 through 67. Chapters 17 and 18 of this text focus on the two forms of hazardous materials that are routinely encountered by all fire code officials: compressed gases, and flammable and combustible liquids.

The requirements in IFC Chapter 50 are used in conjunction with the specific regulations for each class of hazardous materials regulated. For example, consider a building that is storing 300 gallons of an oxidizing liquid. Proper code application would require the enforcement of the requirements in IFC Chapters 50 and 63: Chapter 50 for the general hazardous material provisions and Chapter 63 for the oxidizing liquid provisions. Chapter 50 contains general requirements in Sections 5001 and 5003 that are applicable regardless of the quantity of hazardous materials in storage or use. Section 5004 contains requirements that are applicable when the amount of storage exceeds a certain quantity threshold, known as a maximum allowable quantity per control area (MAQ). Section 5005 specifies requirements that are applicable when the hazardous materials are being used or dispensed, and sets forth specific provisions when the amount of hazardous materials in use exceeds the MAQ. Chapter 63 contains requirements that apply below the MAQ and limit the use of oxidizers in certain occupancies. This material-specific chapter also contains requirements that apply above the MAQ and are in addition to the requirements in Chapter 50. **[Ref. 5001.1, 6301.1.1, 6303.1.1, 6304.1.4]**

Chapter 50 of the IFC sets forth the minimum requirements for storage, use, handling and dispensing of hazardous materials. The terms "storage," "use," "handling" and "dispensing" are defined in the fire code, and understanding them is important to the proper application of the code. These terms describe the "situations" in which hazardous materials occur, and they affect the MAQ. Each class of hazardous material is assigned a MAQ based on its hazards and the relative risk of the material to the building and its occupants. For example, materials in "storage" exist in their original, unopened containers such as those one might find in the paint section of a hardware store. Since the packaging has not been opened and the contents have not mixed with the atmosphere, the code allows a greater quantity to occur within a specific area than if the same product were in "use" or "dispensing" where the material could be spilled, spread or otherwise exposed to an environment where it creates a hazard. Remember that as long as hazardous materials remain in their containers, generally they are easier to control. It is accepted that at some point during the storage of materials, a spill will occur. The product is still considered to be in storage, because the normal situation is storage. The abnormal situation is the spill, so the code

> **Code Essentials**
>
> The Chapter 50 provisions apply to all hazardous materials regulated by the IFC. Chapter 50 is used in conjunction with the requirements for each specific class of hazardous materials regulated in Chapters 51 through 67. Its requirements apply to new and waste hazardous materials. Exemption from the Chapter 50 requirements does not relieve the permit applicant from complying with other applicable I-Code requirements.

FIGURE 16-1 This natural gas pipeline is regulated by the U.S. Department of Transportation – Office of Pipeline Safety and Hazardous Materials. The Federal Pipeline Safety Act preempts the IFC requirements.

Code Essentials

Most materials and uses exempted from Chapter 50 requirements are regulated elsewhere.

For example,

- Off-site transportation of hazardous materials is regulated by the Department of Transportation.
- Mechanical refrigeration is regulated in Section 605.
- Stationary storage battery systems are regulated in Section 1206.2.
- Explosives and fireworks are regulated in Chapter 56.
- Wall-mounted alcohol-based hand rubs are regulated in Section 5705.5.

contains requirements that address the potential incident. The requirements for products in use are more restrictive than for products in storage, since the normal situation is use and the potential for a spill or release is greater.

The requirements in Chapter 50 apply to new (i.e., virgin or unused) and waste chemicals, including their transportation on the site where they are stored, used, handled or dispensed.

Certain processes are exempt from the Chapter 50 requirements. Exemption from Chapter 50 requirements does not relieve the permit applicant from compliance with other IFC provisions and the other adopted codes in the jurisdiction. Exemption also does not relieve the permit applicant from the IFC hazardous materials classification requirements. Off-site transportation is regulated by the U.S. Department of Transportation and is exempt from IFC regulations (Figure 16-1). (Review Chapter 1 for off-site transportation requirements.) [Ref. 5001.1, Exception 4]

The IFC exemptions address specific processes or activities that store or use hazardous materials. Some of the exemptions concern activities that are regulated by other International Code Council (ICC) code provisions and others exempt activities regulated and pre-empted by federal law. For example, mechanical refrigeration systems, which can use refrigerants that are flammable or corrosive (such as anhydrous ammonia), are exempt from the Chapter 50 requirements but are regulated by the requirements in Section 606 and the *International Mechanical Code* (IMC). Stationary storage battery systems using corrosive materials suspended in an aqueous gel are regulated by the requirements in Section 608. These specific provisions offer greater safety when compared to the general and material specific requirements in Chapters 50 through 67. [Ref. 5001.1, Exceptions]

MATERIAL CLASSIFICATION

Before one can apply many of the IFC building safety requirements, the building must be assigned an occupancy classification. A similar exercise is required when applying IFC regulations for storage, handling, dispensing and use of hazardous materials; the hazardous materials must be classified based on their

- Hazard—physical or health hazards
- State—solid, liquid or gas
- Situation—storage or use. If use, then use-open or use-closed.

The IFC classification system recognizes that many hazardous materials represent more than one hazard—all of the hazards of the particular chemical or compound must be addressed. Each hazardous material subject to regulation must be evaluated and assigned its hazard classifications and ratings. [Ref. 5001.1]

Hazardous materials classification can be challenging. Classifying hazardous materials requires a basic understanding of each class of hazardous material regulated by the IFC. Classifications can be complicated when a material is diluted in water, intentionally mixed with other chemicals or its physical form or state is modified. The IFC allows classifications to be assigned using nationally recognized standards, an approved organization or individual, safety data sheets or by other approved means. Safety data sheets (SDS) were formerly known as material safety data sheets (MSDS). The new term is the result of the Globally Harmonized System for Hazard Communication. In 2003, the United Nations adopted the Globally Harmonized System of Classification and Labeling of Chemicals. The intent is to provide a single document containing all the hazard information that is usable and accepted worldwide. The SDS is a result of this process. The fire code official should accept an SDS as equivalent to the traditional MSDS. [Ref. 202, 5001.2.1]

A number of sources are available to assist in classification of hazardous materials. IFC Appendix E contains examples of selected hazardous materials based on the Chapter 50 classification system. Another economical and reliable information source available from ICC is the Hazardous Materials Expert Assistant (HMEX) (Figure 16-2). HMEX is an electronic database cataloging over 8,000 hazardous materials, compounds and product formulations based on the IFC classification criteria. It also contains basic physical data for each material. This program contains material-specific information based on IFC material definitions, incompatibility criteria with other materials, DOT material definitions and SARA Title III (SARA is the federal Superfund Amendments and Reauthorization Act; Title III is the Emergency Planning and Community Right-to-Know Act).

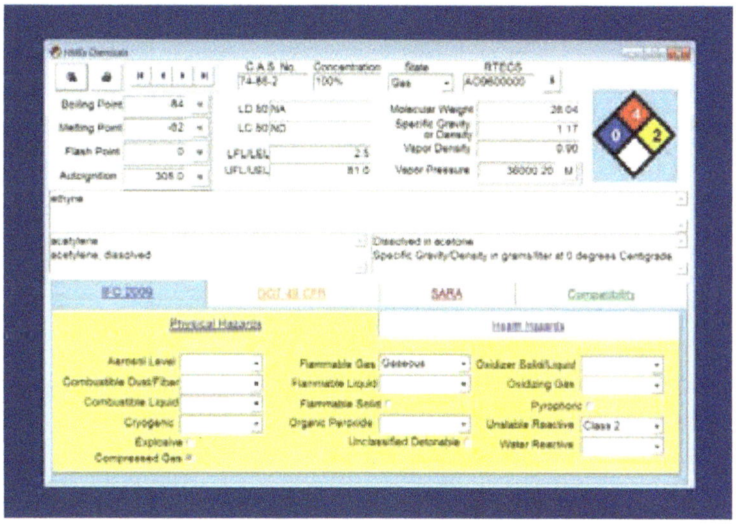

Hazardous materials are commonly manufactured, shipped and consumed as a mixture with some diluent such as water or nitrogen gas. Diluted materials fulfill a market or customer requirement. In other cases, chemicals are not diluted but may be mixed together to serve a particular purpose or application. Mixtures are required to be classified based on their hazards as a whole. [Ref. 5001.2.1]

FIGURE 16-2 HMEX screen capture for acetylene gas

Consider a mixture of sulfuric acid and water. Sulfuric acid is used in the manufacturing of aerospace, automotive, food, electronics, textile, pharmaceutical and plastics products. Like other hazardous materials, its hazards can change based on the amount (sometimes termed as "strength") of sulfuric acid in solution with water. As illustrated in Table 16-1, the hazards of sulfuric acid change when the percentage of acid in water lessens.

TABLE 16-1 Hazard classifications of selected concentrations of sulfuric acid

Concentration (percent volume of sulfuric acid in water)	IFC hazard classification	Common uses
98%	Class 2 Water Reactive, Corrosive, Toxic, Class 1 Oxidizer	Fertilizer manufacturing; cleaning of semiconductor wafers
32–38%	Class 1 Water Reactive, Corrosive	Electrolyte in wet-cell automobile batteries
2.5%	Nonhazardous	Cutting an onion releases an amino acid known as a sulfoxide. When the sulfoxide mixes with the water in a human eye, it forms a weak solution of sulfuric acid.

The IFC hazardous materials classification system is based on the classification system used by the U.S. Department of Labor, Occupational Safety and Health Administration (OSHA). The OSHA criterion is used because its classification method focuses on workplace safety, which matches the intent and scope of the fire code. The IFC classification system is generally not consistent with the U.S. DOT or Environmental Protection Agency classification systems. These and other federal agencies have different classification criterion because of the differences in, and the goals of, their regulations. **[Ref. 5001.2.2]**

Hazardous materials classified using the IFC scheme are physical hazards, health hazards or a combination of both. A single hazardous material can present one or more physical and health hazards. Physical hazard materials burn, accelerate burning, or could either detonate or deflagrate. Many of the physical hazard materials that burn can produce a deflagration under certain conditions, such as when heated or if they are pressurized. Conversely, certain classes of oxidizers and most classes of explosives are capable of producing detonations. Oxidizers are the only physical hazard that can chemically accelerate burning. **[Ref. 5001.2.2.1]**

Many physical hazard materials are assigned a class designation. The class designation is a relative estimation of the potential outcome if the hazardous material is improperly stored or used. The class designation provides an indication of the ease or difficulty of igniting the material or initiating an uncontrolled chemical reaction and the potential outcomes if the material is improperly stored or handled.

Different physical hazard materials can have a different number of hazard classes. The classification numbering convention is not consistent among all the physical hazard materials. For example, a Class 4 oxidizer is more hazardous than a Class 1 oxidizer. The opposite is true for organic peroxides where a Class I organic peroxide is more dangerous than a Class IV organic peroxide. This anomaly in the classification criteria is a result of how hazardous materials classification criteria in consensus safety standards were developed before they were incorporated into the model codes (Figure 16-3).

Arabic numeral	Potential	Roman numeral
4	GREATEST POTENTIAL	I
3		II
2		III
		IV
1	LEAST POTENTIAL	V

FIGURE 16-3 Physical hazard numbering convention

IFC Chapter 50 regulates the following hazardous materials that are a physical hazard:
1. Explosives and blasting agents
2. Combustible liquids
3. Flammable solids, liquids and gases
4. Organic peroxide solids or liquids
5. Oxidizer, solids or liquids
6. Oxidizing gases
7. Pyrophoric solids, liquids or gases
8. Unstable (reactive) solids, liquids or gases
9. Water-reactive solids or liquids
10. Cryogenic fluids. [Ref. 5001.2.2.1]

The 14th-century alchemist Paracelsus, considered by some to be the father of toxicology, wrote: "[a]ll things are poison and nothing is without poison, only the dose permits something not to be poisonous." Applying this philosophy, materials such as water and dirt are poisons, and annually people die because of drowning or being trapped when an excavated trench collapses. The IFC regulates health hazard materials that cause death, injury or incapacitation if an individual has a single, brief exposure to the hazardous material.

The ratings for health hazard materials are based on analytical and laboratory testing of living organisms and the three routes that the human body can be exposed to hazardous materials: inhalation, absorption and ingestion. The results of these tests determine if a health hazard material is classified as
1. Highly toxic solid, liquid or gas
2. Toxic solid, liquid or gas
3. Corrosive solid, liquid or gas. [Ref. 5001.2.2.2]

> **Code Essentials**
>
> Physical hazards are capable of burning, accelerating burning, deflagrations or detonations. Many of the physical hazard materials are assigned a class designation, which is a relative estimation of the outcome if the hazardous material is not correctly stored and handled.

> **Code Essentials**
>
> LC_{50} values measure the toxic effects of hazardous materials whose route of exposure is inhalation. LD_{50} values measure toxic effects of hazardous materials whose route of exposure is through skin absorption or ingestion.

The distinction among health hazard materials as being highly toxic or toxic is based on reproducible results from tests supervised by toxicologists. These tests, in conjunction with other scientific literature, are used to establish the values known as LC_{50} and LD_{50} (Table 16-2). The term LC_{50} is a measurement of the lethal concentration of a hazardous material that kills 50 percent of the animals tested. LC_{50} values are measurements of toxins that are inhaled and the measurement is expressed in milligrams of toxin/liter of air (mg/L) or in parts per million or billion (PPM or PPB) of the contaminant in air. LD_{50} is the measurement of the amount of a toxin that kills 50 percent of the test animals when exposed to the chemical via absorption through skin or by ingestion. LD_{50} values are expressed in milligrams of toxin/kilograms of animal weight (mg/Kg). The lower the LD_{50} or LC_{50} value, the more powerful the toxin because it requires less of the material to cause 50 percent of the test animals to die. The IFC criterion for classifying these materials is based on specific animal species of a certain body weight. [Ref. 202]

TABLE 16-2 IFC classification criterion for highly toxic and toxic materials

IFC classification	Exposure route		
	Inhalation toxicity threshold	Absorption toxicity threshold	Ingestion toxicity threshold
Highly Toxic	200 ppm or less; 2 mg/L or less	200 mg/kg or less	50 mg/kg or less
Toxic	Greater than 200 ppm but not more than 2,000 ppm; greater than 2 mg/L but not more than 20 mg/L	Greater than 200 mg/kg but not more than 1,000 mg/kg	Greater than 50 mg/kg but not more than 500 mg/kg

mg/kg = milligram/kilogram of body weight
mg/L = milligram/liter of mist, fume, or dust
ppm = parts per million of gas or vapor

FIGURE 16-4 DOT corrosive label

Materials that cause visible destruction or irreversible alterations in living tissues by a chemical action are classified as corrosive. Corrosives can be a solid, liquid or gas. The IFC definition of "corrosive" specifically excludes materials that chemically react when they contact an inanimate object, such as metals. One example is ferric chloride used in water treatment and metal etching. Ferric chloride aggressively attacks most metals yet will not harm human skin. A hazardous material is classified as a corrosive based on the test method specified by the U.S. DOT. Because the IFC references the DOT test method, a hazardous material with a corrosive label or placard is generally assigned the same IFC hazard classification except for materials that do not cause irreversible damage to human skin (Figure 16-4).

EXAMPLE

Acrolein is used in the manufacturing of a number of plastics. Acrolein has a flash point temperature of -15°F and a boiling point temperature of 127°F. The material has an absorption LD_{50} value of 200 mg/kg and an ingestion LD_{50} value of 26 mg/kg. Based on the definitions and classifications for flammable and combustible liquids and highly toxic and toxic materials in IFC Section 202, what is its classification?

ANSWER

Acrolein is a Class I-B flammable and highly toxic liquid.

HAZARDOUS MATERIALS REPORTING

The requirement for an IFC operational permit to store, handle, use and dispense a hazardous material is based on the permit applicant reporting the hazardous material classification, physical state, the method it will be used and the amount of material. When required by the fire code official, the applicant for a hazardous materials construction or operational permit may be required to submit either a Hazardous Materials Management Plan (HMMP) or a Hazardous Materials Inventory Statement (HMIS). Example formats for HMMP and HMIS submittal documents are contained in IFC Appendix H (Figure 16-5).

Code Essentials

"Control area" is a space within the building where a limited quantity (maximum allowable quantity per control area or MAQ) of hazardous materials can be stored or used without affecting the occupancy classification.

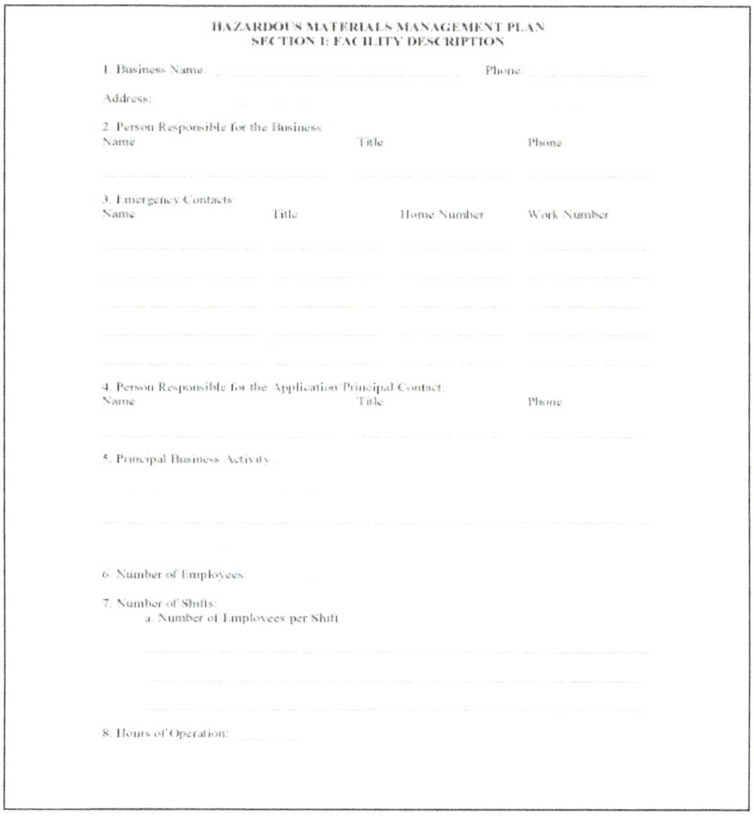

HAZARDOUS MATERIALS MANAGEMENT PLAN
SECTION I: FACILITY DESCRIPTION

1. Business Name: Phone:

Address:

2. Person Responsible for the Business
Name Title Phone

3. Emergency Contacts:
Name Title Home Number Work Number

4. Person Responsible for the Application/Principal Contact:
Name Title Phone

5. Principal Business Activity

6. Number of Employees:

7. Number of Shifts:
a. Number of Employees per Shift

8. Hours of Operation:

FIGURE 16-5 Sample Hazardous Materials Management Plan cover page

The HMMP documents basic information so emergency responders understand the construction and access routes for a building or premises, such as emergency exits that can be used for access, the physical and health hazards of hazardous materials stored and used within particular areas, where emergency responders will meet the fire department liaison and the location of all above-ground and underground tanks, sumps, vaults or any other below-grade processes. This information is extremely beneficial to emergency responders because it identifies locations where confined space entries may need to be performed. The HMMP also identifies building areas constructed as control areas and Group H occupancies. [Ref. 5001.5.1]

An HMIS contains information beneficial to building and fire code officials, plans examiners and inspectors attempting to determine that the amount of hazardous materials in storage and use complies with the IFC and IBC requirements (Figure 16-6). An HMIS documents the product's name, its chemical constituents along with their respective Chemical Abstracts Service (CAS) number, the volume of containers or tanks, the product's hazard classification and the amount in storage, use-open and use-closed systems. This information is beneficial for plans examiners to confirm the occupancy classification, if the mechanical ventilation system should comply with the IMC requirements for hazardous exhaust systems and to verify that the design of the processes complies with the applicable requirements of the IFC. [Ref. 5001.5.2]

HAZARDOUS MATERIALS INVENTORY STATEMENT (HMIS)
SUMMARY REPORT
STORAGE CONDITIONS

Hazard Class	Hazard Subclass (Abbrev)	Inventory Amount			Maximum Allowable Quantity		
		Solid (lb.)	Liquid (gal.)	Gas (cu. ft.)	Solid (lb.)	Liquid (gal.)	Gas (cu. ft.)
Combustible Liquid	C-II		25			120	
	C-IIIA					330	
	C-IIIB		160			13,200	
Corrosive	Cor		15			500	
Flammable Gas	FLG			150			1,000
Flammable Liquid	F-IA					30	
	F-IB & F-IC		15			120	
	Combined F-IA, F-IB & F-IC		15			120	
Oxidizer	OX			600			1,500
Toxic	T		15			50	
Unstable Reactive	UR4						10
	UR3						50
	UR2			150			750
	UR1						NL
Water Reactive	WR3					0.5	
	WR2					5	
	WR1		15			NL	

INVENTORY REPORT

Product Name (Components)	CAS Number	Location	Container >55 gallon	Hazard Class 1	Hazard Class 2	Hazard Class 3	Stored (lbs.)	Stored (gal.)	Stored (cu.ft.)	Use-closed (lbs.)	Use-closed (gal.)	Use-closed (cu.ft.)	Use-open (lbs.)	Use-open (gal.)	Use-open (cu.ft.)
Acetylene Gas	74-86-2	Control Area 1		FLG	UR2				150						
Diesel	68476-34-6	Control Area 2		C-II				25							
Gasoline, Unleaded	8006-61-9	Control Area 1		F-IB				15							
Motor Oil 10W/40	64762-54-7	Control Area 1	Yes	C-IIIB				105							
Oxygen Gas	7782-44-7	Control Area 1		OX					600						
Sulfuric Acid	7664-93-9	Control Area 2		Cor	T	WR1		15							
Transmission Fluid	64742-65-0	Control Area 2	Yes	C-IIIB				55							

FIGURE 16-6 Sample Hazardous Materials Inventory Statement

The IFC generally requires any accidental hazardous material release be reported to the fire code official. "A hazardous material release or emission in a manner that does not conform to the provisions of the IFC or applicable health and safety regulation" is the definition of an "unauthorized discharge." When an unauthorized discharge of hazardous materials results from a container failure, the fire code official is authorized to require its repair or removal from service. Those responsible for the unauthorized discharge are responsible for the costs associated to clean up any hazardous material release. If fire department resources such as protective clothing, firefighting foam or personnel costs are expended to respond and mitigate an unauthorized discharge, the fire code official is authorized to recover the costs (Figure 16-7). [Ref. 202, 5003.3.1]

FIGURE 16-7 The fire department is authorized to recover the costs associated with the response to this unauthorized discharge and subsequent flammable liquid fire. *(Courtesy of Kern County [CA] Fire Department)*

STORAGE AND USE

The situation of a hazardous material is an important part of the IFC and IBC hazardous materials regulations. The IFC recognizes hazardous materials can be in one of the following three situations:
- Storage
- Use-closed system
- Use-open system.

For hazardous materials, "storage" is defined as "the keeping, retention, or leaving of hazardous materials in closed containers, tanks, cylinders, or similar vessels; or vessels supplying operations through closed connections to the vessel." A stored hazardous material is essentially static and dormant and is generally awaiting use. IFC Section 5004 requirements are applicable to storage exceeding MAQ and are based on the material's classification and physical state. [Ref. 202, 5004.1]

A material that is "placed into action" is defined by the IFC as being in "use." Use occurs when a material's stored energy is released, kinetic energy is introduced, mixtures are created or gravity is used to facilitate its movement. IFC Section 5005 has requirements that are applicable to use where the quantity exceeds the MAQ and are based on the material's classification and physical state. The use of hazardous materials is regulated because placing the material into action introduces the possibility of an unauthorized discharge and therefore the piping or process systems must be reviewed and approved by the fire code official. Use can be within a closed or open system. [Ref. 202, 5005.1]

> **Code Essentials**
>
> The IFC categorizes the situation of hazardous materials into either storage or use. The situation must be determined when applying the IFC hazardous material provisions. If the material is in use, this can occur in either an open system or a closed system. Only solids and liquids can be used in an open system.

A use-closed system is the "use of a solid or liquid hazardous material involving a closed vessel or system that remains closed during normal operations where vapors emitted by the product are not liberated outside of the vessel or system and the product is not exposed to the atmosphere during normal operations; and all uses of compressed gases. Examples of closed systems for solids and liquids include product conveyed through a piping system into a closed vessel, system or piece of equipment." [Ref. 202]

A use-open system involves solid or liquid hazardous materials. It does not include compressed gases. Compressed gases are not regulated as an open system, even though they may be released to the atmosphere at the point of use, such as with oxygen-acetylene welding. A use-open system is "the use of a solid or liquid hazardous material involving a vessel or system that is continuously open to the atmosphere during normal operations and where vapors are liberated, or the product is exposed to the atmosphere during normal operations." Open systems for solids and liquids include dispensing from or into open beakers or containers, dip tanks and plating tank operations (Figure 16-8). [Ref. 202]

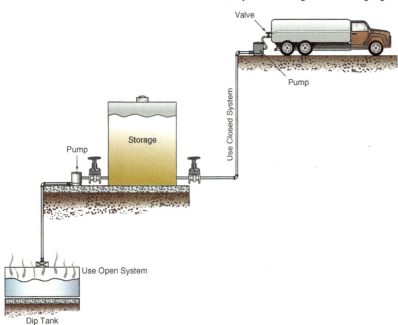

FIGURE 16-8 This graphic depicts the three situations for hazardous materials: use-closed (product delivery through fixed connections), storage (storage tank) and use-open systems (open-top dip tank).

When determining whether the situation of a material is use-open or use-closed, the fire code official must consider the normal operation of the vessel or system. The key distinction between use-open and use-closed is that use-open is "continuously open to the atmosphere during normal operations," while use-closed is "closed during normal operations." Consider a tank that is used to mix and blend various materials. All the materials are piped into the blending vessel, and the completed mixture is piped to a storage tank. This is a use-closed system. However, during the blending process a cover is opened to obtain a sample of the mixture for testing, and then the cover is closed. The normal operation is closed, so this is still considered use-closed. Similarly, at some time the vessel will need to be opened for maintenance and repair. Even though this tank is now open to the atmosphere and could release vapors into the room, it is considered a use-closed system because it is "closed during normal operations." It is opened and releases vapors to the atmosphere; however, it is not "continuously open."

MAXIMUM ALLOWABLE QUANTITY PER CONTROL AREA

The amount of hazardous material allowed to be stored and used inside of a building is based on the material's hazard classification, situation, and physical state. The maximum allowable quantity per control area (MAQ) is exactly as the term implies: it is the maximum amount of a class of hazardous material that is permitted within a control area without requiring the building to be classified as a Group H (hazardous) occupancy. The MAQ values were developed based on the relative hazards associated with each regulated class of hazardous materials.

The MAQ values in the IFC are found in Tables 5003.1.1(1) through (4). Table 5003.1.1(1) sets forth the MAQ for physical hazard materials inside of buildings, and Table 5003.1.1(2) sets forth the health hazard materials MAQ inside of buildings. There are two more tables for outdoor storage and use. [In the context of this book, Table 16-3 represents Table 5003.1.1(1), and it is to explain the concepts of MAQ and control areas. Readers are encouraged to review the 2018 IFC to understand and apply the concept of MAQ and control areas for health hazard materials and outdoor control areas.]

Notice that the table title provides the description "Maximum Allowable Quantity *per Control Area*" (emphasis added). This is an important distinction for the code official and the regulated community. A control area is a space inside or outside of a building where the quantity of hazardous materials stored, used, handled or dispensed does not exceed the MAQ. With the addition of two or more control areas, an occupant can increase the amount of hazardous materials in a building without having to reclassify it as a Group H occupancy.

The IFC MAQ tables are divided into five major headings:
- The classification of the hazardous material
- The occupancy classification of the building when the MAQ is exceeded
- Storage
- Use-closed system
- Use-open system. [Ref. Table 5003.1.1(1)]

Under the headings for storage and use, each group of table columns is divided by the three physical states and the unit of measurement. The only exception to this is under use-open systems, where gases are not listed since gases are always considered use-closed. Depending on the material classification, the MAQ for liquids is based on the volume (gallons) or weight (pounds) of the material. In other words, a Class III organic peroxide has a MAQ of 125 pounds if it is a solid or liquid.

TABLE 16-3 Maximum allowable quantity per control area of hazardous materials posing a physical hazard[a, j, m, n, p] [IFC Table 5003.1.1(1)]

Material	Class	Group when the maximum allowable quantity is exceeded	Storage[b] Solid pounds (cubic feet)	Storage[b] Liquid gallons (pounds)	Storage[b] Gas cubic feet at NTP	Use-closed systems[b] Solid pounds (cubic feet)	Use-closed systems[b] Liquid gallons (pounds)	Use-closed systems[b] Gas cubic feet at NTP	Use-open systems[b] Solid pounds (cubic feet)	Use-open systems[b] Liquid gallons (pounds)
Combustible dust	NA	H-2	Note q	NA	NA	Note q	NA	NA	Note q	NA
Combustible fiber[o]	Loose baled[o]	H-3	(100) (1,000)	NA	NA	(100) (1,000)	NA	NA	(20) (200)	NA
Combustible liquid[c, i]	II / IIIA / IIIB	H-2 or H-3 / H-2 or H-3 / NA	NA	120[d,e] / 330[d,e] / 13,200[e,f]	NA	NA	120[d] / 330[d] / 13,200[f]	NA	NA	30[d] / 80[d] / 3,300[f]
Cryogenic flammable	NA	H-2	NA	45[d]	NA	NA	45[d]	NA	NA	10[d]
Cryogenic inert	NA	NA	NA	NA	NL	NA	NA	NL	NA	NA
Cryogenic oxidizing	NA	H-3	NA	45[d]	NA	NA	45[d]	NA	NA	10[d]
Explosives	Division 1.1 / Division 1.2 / Division 1.3 / Division 1.4 / Division 1.4G / Division 1.5 / Division 1.6	H-1 / H-1 / H-1 or H-2 / H-3 / H-3 / H-1 / H-1	1[e,g] / 1[e,g] / 5[e,g] / 50[e,g] / 125[e,i] / 1[e,g] / 1[e,g]	(1)[e,g] / (1)[e,g] / (5)[e,g] / (50)[e,g] / NA / (1)[e,g] / NA	NA	0.25[g] / 0.25[g] / 1[g] / 50[g] / NA / 0.25[g] / NA	(0.25)[g] / (0.25)[g] / (1)[g] / (50)[g] / NA / (0.25)[g] / NA	NA	0.25[g] / 0.25[g] / 1[g] / NA / NA / 0.25[g] / NA	(0.25)[g] / (0.25)[g] / (1)[g] / NA / NA / (0.25)[g] / NA
Flammable gas	Gaseous / Liquefied	H-2	NA	NA / (150)[d,e]	1,000[d,e] / NA	NA	NA / (150)[d,e]	1,000[d,e] / NA	NA	NA
Flammable liquid[c]	IA / IB and IC	H-2 or H-3	NA	30[d,e] / 120[d,e]	NA	NA	30[d] / 120[d]	NA	NA	10[d] / 30[d]
Flammable liquid, combination (IA, IB, IC)	NA	H-2 or H-3	NA	120[d,e,h]	NA	NA	120[d,h]	NA	NA	30[d,h]
Flammable solid	NA	H-3	125[d,e]	NA	NA	125[d]	NA	NA	25[d]	NA
Inert gas	Gaseous / Liquefied	NA / NA	NA / NA	NA / NA	NL / NL	NA / NA	NA / NA	NL / NL	NA / NA	NA / NA

(Continued)

Maximum Allowable Quantity Per Control Area

TABLE 16-3 Maximum allowable quantity per control area of hazardous materials posing a physical hazard[a,j,m,n,p] [IFC Table 5003.1.1(1)]—Cont'd

Material	Class	Group when the maximum allowable quantity is exceeded	Storage[b]			Use-closed systems[b]			Use-open systems[b]	
			Solid pounds (cubic feet)	Liquid gallons (pounds)	Gas (cubic feet at NTP)	Solid pounds (cubic feet)	Liquid gallons (pounds)	Gas (cubic feet at NTP)	Solid pounds (cubic feet)	Liquid gallons (pounds)
Organic peroxide	UD	H-1	1[e,g]	(1)[e,g]	NA	0.25[g]	(0.25)[g]	NA	0.25[g]	(0.25)[g]
	I	H-2	5[d]	(5)[d]		1[d]	(1)[d]		1[d]	(1)[d]
	II	H-3	50[d]	(50)[d]		50[d]	(50)[d]		10[d]	(10)[d]
	III	H-3	125[d]	(125)[d]		125[d]	(125)[d]		25[d]	(25)[d]
	IV	NA	NL	NL		NL	NL		NL	NL
	V	NA	NL	NL		NL	NL		NL	NL
Oxidizer	4	H-1	1[e,g]	(1)[e,g]	NA	0.25[g]	(0.25)[g]	NA	0.25[g]	(0.25)[g]
	3[k]	H-2 or H-3	10[d,e]	(10)[d,e]		2[d]	(2)[d]		2[d]	(2)[d]
	2	H-3	250[d,e]	(250)[d,e]		250[d]	(250)[d]		50[d]	(50)[d]
	1	NA	4,000[e,f]	(4,000)[e,f]		4,000[f]	(4,000)[f]		1,000[f]	(1,000)[f]
Oxidizing gas	Gaseous	H-3	NA	NA	1,500[d,e]	NA	NA	1,500[d,e]	NA	NA
	Liquefied			(150)[d,e]	NA		(150)[d,e]	NA		
Pyrophoric	NA	H-2	4[e,g]	(4)[e,g]	50[e,g]	1[g]	(1)[g]	10[e,g]	0	0
Unstable (reactive)	4	H-1	1[e,g]	(1)[e,g]	10[e,g]	0.25[g]	(0.25)[g]	2[e,g]	0.25[g]	(0.25)[g]
	3	H-1 or H-2	5[d,e]	(5)[d,e]	50[d,e]	1[d]	(1)[d]	10[d,e]	1[d]	(1)[d]
	2	H-3	50[d,e]	(50)[d,e]	750[d,e]	50[d]	(50)[d]	750[d,e]	10[d]	(10)[d]
	1	NA	NL	NL	NL	NL	NL	NL	NL	NL
Water reactive	3	H-2	5[d,e]	(5)[d,e]	NA	5[d]	(5)[d]	NA	1[d]	(1)[d]
	2	H-3	50[d,e]	(50)[d,e]		50[d]	(50)[d]		10[d]	(10)[d]
	1	NA	NL	NL		NL	NL		NL	NL

For SI: 1 cubic foot = 0.02832 m^3, 1 pound = 0.454 kg, 1 gallon = 3.785 L.
NA = Not Applicable; NL = Not Limited; UD = Unclassified Detonable.

a. For use of control areas, see Section 5003.8.3.
b. The aggregate quantity in use and storage shall not exceed the quantity listed for storage.
c. The quantities of alcoholic beverages in retail and wholesale sales occupancies shall not be limited providing the liquids are packaged in individual containers not exceeding 1.3 gallons. In retail and wholesale sales occupancies, the quantities of medicines, foodstuffs or consumer products and cosmetics containing not more than 50 percent by volume of water-miscible liquids with the remainder of the solutions not being flammable shall not be limited, provided that such materials are packaged in individual containers not exceeding 1.3 gallons.
d. Maximum allowable quantities shall be increased 100 percent in buildings equipped throughout with an approved automatic sprinkler system in accordance with Section 903.3.1.1. Where Note e also applies, the increase for both notes shall be applied accumulatively.
e. Maximum allowable quantities shall be increased 100 percent when stored in approved storage cabinets, day boxes, gas cabinets, gas rooms, exhausted enclosures or listed safety cans in accordance with Section 5003.9.10. Where Note d also applies, the increase for both notes shall be applied accumulatively.
f. Quantities shall not be limited in a building equipped throughout with an approved automatic sprinkler system in accordance with Section 903.3.1.1.
g. Allowed only in buildings equipped throughout with an approved automatic sprinkler system.
h. Containing not more than the maximum allowable quantity per control area of Class IA, Class IB or Class IC flammable liquids.
i. The maximum allowable quantity shall not apply to fuel oil storage complying with Section 603.3.2.
j. Quantities in parenthesis indicate quantity units in parenthesis at the head of each column.

(Continued)

TABLE 16-3 maximum allowable quantity per control area of hazardous materials posing a physical hazard[a,j,m,n,p] [IFC Table 5003.1.1(1)] —Cont'd

k. A maximum quantity of 220 pounds of solid or 22 gallons of liquid Class 3 oxidizers is allowed when such materials are necessary for maintenance purposes, operation or sanitation of equipment where the storage containers and the manner of storage are approved.

l. Net weight of pyrotechnic composition of the fireworks. Where the net weight of the pyrotechnic composition of the fireworks is not known, 25 percent of the gross weight of the fireworks including packaging shall be used.

m. For gallons of liquids, divide the amount in pounds by 10 in accordance with Section 5003.1.2.

n. For storage and display quantities in Group M and storage quantities in Group S occupancies complying with Section 5003.11, see Table 5003.11.1.

o. Densely-packed baled cotton that complies with the packing requirements of ISO 8115 shall not be included in this material class.

p. The following shall not be included in determining the maximum allowable quantities:
  1. Liquid or gaseous fuel in fuel tanks on vehicles.
  2. Liquid or gaseous fuel in fuel tanks on motorized equipment operated in accordance with this code.
  3. Gaseous fuels in piping systems and fixed appliances regulated by the *International Fuel Gas Code*.
  4. Liquid fuels in piping systems and fixed appliances, regulated by the *International Mechanical Code*.
  5. Alcohol-based hand rubs classified as Class I or II liquids in dispensers that are installed in accordance with Sections 5705.5 and 5705.5.1. The location of the alcohol-based hand rub (ABHR) dispensers shall be provided in the construction documents.

q. Where manufactured, generated or used in such a manner that the concentration and conditions create a fire or explosion hazard based on information prepared in accordance with Section 104.7.2.

Both solid and liquid states are measured in pounds because many processes that use this particular class of hazardous material measure the quantity based on weight (pounds) rather than volume (gallons). In instances where a material is reported in gallons and is regulated by weight, Footnote m and Section 5003.1.2 specify a conversion factor of 10 pounds/gallon unless the liquid's density or specific gravity is provided. Using the safety data sheet (SDS), the actual conversion could be calculated, but the conversion of 10:1 can be used and is a conservative value. To differentiate the correct unit of measurement, MAQ values that are parenthetically cited are based on the measurement units referenced at the top of the column in the "Material" row of Tables 5003.1.1(1) (Table 16-3) and 5003.1.1(2). [Ref. 5003.1.2]

The IFC is more restrictive in its requirements for materials in use. For materials used in a use-open system, the code establishes a much lower MAQ when compared to materials in use-closed systems (Figure 16-9). Consider a Class IC flammable liquid. Its use-closed MAQ is 120 gallons versus a use-open system MAQ of 30 gallons. Similar reductions are presented for other physical and health hazard materials. The reason for the reduction is use-open systems present a greater number of hazards when compared to use-closed systems because they are constantly open to the atmosphere of a room and vapors are liberated, which increases the likelihood of the vapors being ignited or causing a chemical exposure. If multiple materials are in use-open systems, this increases the potential of an accidental mixing of incompatible hazardous materials. Accordingly, the code reduces the MAQ for materials in use-open systems. [Ref. Tables 5003.1.1(1), 5003.1.1(2)]

FIGURE 16-9 This open dip tank of a combustible liquid is an example of a use-open system.

When using Tables 5003.1.1(1) through (4) for determining the MAQ in one or more control areas, it is important to review and apply the various footnotes. The footnotes provide supplemental requirements for the proper application of these tables. Footnote b in Tables 5003.1.1(1) through (4) stipulates that the aggregate quantity in storage and in use cannot exceed the MAQ listed for storage. For example, the storage MAQ for a Class IIIA combustible liquid is 330 gallons and its use-open system MAQ is 80 gallons. Assume that an industrial process requires the open use of a Class IIIA combustible liquid and the dip tank volume is 80 gallons. The proper application of this footnote would result in 250 gallons in storage and 80 gallons in a use-open system (Figure 16-10). [Ref. Table 5003.1.1(1)]

Certain hazardous materials present such a high fire or explosion risk that the IFC requires the installation of an automatic sprinkler system throughout

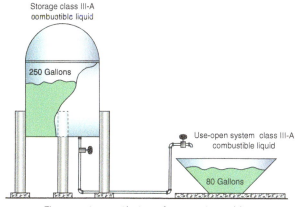

FIGURE 16-10 250 gallons in storage and 80 gallons in use-open results in an aggregate of 330 gallons of Class IIIA combustible liquid.

FIGURE 16-11 This gas cabinet was involved in a fire resulting from a release of silane. Silane is a pyrophoric and flammable gas. Because of the hazards with silane, the room and gas cabinet were required by the IFC to be protected by an approved automatic sprinkler system, which effectively controlled the fire.

FIGURE 16-12 Installation of an approved automatic sprinkler system increases the MAQ by 100 percent for many hazardous materials.

FIGURE 16-13 Storage in approved cabinets is a permitted method of increasing the MAQ of many hazardous materials.

the building. Footnote g in Table 5003.1.1(1) only permits the storage of explosive materials, pyrophorics and certain classes of organic peroxides, oxidizers and unstable (reactive) materials inside buildings when they are protected by an approved automatic sprinkler system (Figure 16-11). The automatic sprinkler system must be installed in accordance with the requirements in Section 903.3.1.1. **[Ref. Table 5003.1.1(1), Footnote g]**

The IFC allows the amount of hazardous materials in a control area to be increased for most physical hazard and all health hazard materials when a building is protected throughout by an approved automatic sprinkler system that complies with the requirements in Section 903.3.1.1. Footnote d in Tables 5003.1.1(1) and (2) permit a 100-percent quantity increase when automatic sprinkler protection is provided throughout the building (Figure 16-12). This increase is allowed because of the reliability of an approved automatic sprinkler system to control a fire. For example, the MAQ for a solid Class 2 Oxidizer is 250 pounds in a nonsprinklered building. If the building is protected by an approved sprinkler system, the MAQ is increased to 500 pounds. Two items with regard to the proper application of this footnote are important:

1. the automatic sprinkler system must comply with Section 903.3.1.1 which requires a system designed to NFPA 13, and
2. the entire building must be protected with the sprinkler system, not just the room or fire area where the hazardous materials are located. **[Ref. Table 5003.1.1(1), Table 5003.1.1(2)]**

The IFC also allows the quantity to be increased when hazardous materials are located in approved storage cabinets, day boxes, gas cabinets, gas rooms, exhausted enclosures or listed safety cans (Figure 16-13). Footnote e in Tables 5003.1.1(1) and (2) allows a 100 percent increase of the MAQ when one of these storage methods is used. Each indicated method protects the hazardous material by providing a fire-resistive enclosure or second means of containment for the stored product. Doors to the enclosures are generally required to be self-closing and self-latching. When used in conjunction with an automatic sprinkler system, the MAQ can be increased by 400 percent. For example, the MAQ for a Class 2 Unstable (Reactive) liquid is 50 pounds. If the building is protected by an automatic sprinkler system complying with the requirements in Section 903.3.1.1, the MAQ is increased by 100 percent to 100 pounds. If all of the liquid is stored in an approved hazardous materials storage cabinet, the MAQ is increased again by 100 percent to 200 pounds. **[Ref. Table 5003.1.1(1), Footnotes d and e]**

CONTROL AREAS

For the purpose of regulating hazardous materials, all buildings and facilities are considered to be a control area. A control area is a "space inside a building or an outdoor facility where the quantity of hazardous materials stored, used, handled or dispensed does not exceed the MAQ." A building can have more than one control area. The number of control areas and the MAQ of hazardous materials in each control area vary based on the floor level where the control areas are located. **[Ref. 5003.8.3]**

When a building is divided into different spaces for the storage and use of hazardous materials by fire-resistive construction, each space is classified as a control area. Control areas are required to be constructed in accordance with the IBC (Figure 16-14).

The number of control areas allowed inside of a building and the amount of hazardous material that can be stored or used depends on the location of the storage and use in relation to the grade plane, or ground floor level of the building. In a one-story building, four control areas are permitted, each containing up to the MAQ for each hazard class of the material. Footnote a in Table 5003.8.3.2 (Table 16-4) allows the application of all of the quantity increases permitted in Tables 5003.1.1(1) and 5003.1.1(2), such as the installation of an automatic sprinkler system throughout the building that complies with Section 903.3.1.1 and the use of hazardous material storage cabinets, gas cabinets, day boxes and exhausted enclosures. **[Ref. 5003.8.3.1]**

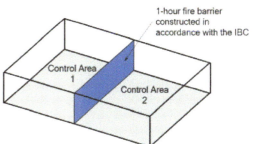

FIGURE 16-14 Control areas in a 1-story building require a minimum 1-hour separation constructed in accordance with the IBC.

TABLE 16-4 Design and number of control areas (IFC Table 5003.8.3.2)

Story		Percentage of the maximum allowable quantity per control area[a]	Number of control areas per story	Fire-resistance rating for fire barriers in hours[b]
Above grade plane	Higher than 9	5	1	2
	7–9	5	2	2
	6	12.5	2	2
	5	12.5	2	2
	4	12.5	2	2
	3	50	2	1
	2	75	3	1
	1	100	4	1
Below grade plane	1	75	3	1
	2	50	2	1
	Lower than 2	Not allowed	Not allowed	Not allowed

a. Percentages shall be of the maximum allowable quantity per control area shown in Tables 5003.1.1(1) and 5003.1.1(2), with all increases allowed in the footnotes to those tables.
b. Separation shall include fire barriers and horizontal assemblies as necessary to provide separation from other portions of the building.

QUESTION

A two-story Group F-1 occupancy houses a manufacturer of specialized optics. The manufacturing process requires the storage and use of a liquid hazardous material classified as a Class II organic peroxide. The building is protected throughout by an approved automatic sprinkler system in accordance with IFC Section 903.3.1.1.

The company wants to know the maximum number control areas that are allowed on each story, the MAQ of organic peroxide permitted in each control area, and the total volume of organic peroxide that can be stored inside the building.

ANSWER

Application of Table 5003.8.3.2 (see Table 16-4) allows up to seven control areas in the building—four control areas at the grade plane level and three control areas on the second floor. The MAQ on the first floor is limited to 100 pounds in each of the four control areas using the requirements in Table 5003.1.1(1), including Footnote d. The MAQ on the second floor is limited to 75 pounds in each of the three control areas. The total volume of Class II Organic Peroxide allowed in the building is 625 pounds, which must be stored so the MAQ per each control area is not exceeded (see Figure 16-15).

Code Essentials

The IFC limits the number of control areas inside a building and the MAQ of hazardous materials in each control area. As the height of the control area above the grade plane level increases, the number of control areas and the MAQ on each floor is reduced. Control areas located four or more stories above the grade plane require separation using minimum 2-hour fire-resistance-rated fire barriers and horizontal assemblies. •

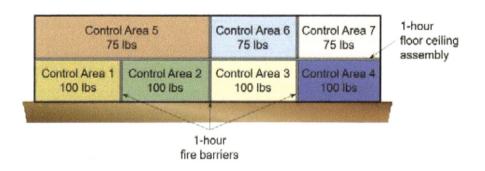

FIGURE 16-15 Example: seven control areas in a two-story building

Construction of fire walls, fire barriers and horizontal assemblies that separate buildings into two or more control areas must comply with the IBC. The IBC requires the construction of fire-resistance-rated assemblies be approved by the building code official. The required fire-resistance rating of the fire barriers separating stories into multiple control areas is specified in Table 5003.8.2. As the location of the control area in relation to the grade plane increases, the fire-resistance rating of walls also increases. When control areas are constructed four or more stories above the grade plane, the fire-resistance rating of fire barriers increases to two hours. Floor assemblies are required to be of at least 2-hour fire-resistance-rated construction. The IBC requires that the structural members supporting a fire-resistance-rated floor/ceiling assembly have at least the same fire-resistance rating as the floor/ceiling assembly they support. Therefore, the supporting structural elements must be at least 2-hour fire-resistance rated to support a control area above. In buildings of Type IIA, IIIA or VA construction three stories or less in height with control areas constructed using fire-resistance-rated assemblies, the IBC and IFC permit the fire-resistance rating of horizontal assemblies to be reduced to 1 hour when the building is protected throughout by an approved automatic sprinkler system complying with the requirements in Section 903.3.1.1 (Figure 16-16). [Ref. 5003.8.3.4]

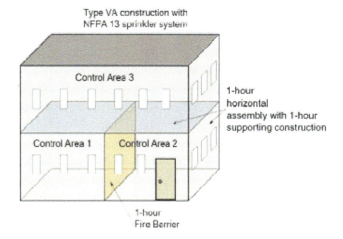

FIGURE 16-16 The fire-resistance rating of floor areas supporting control areas in some buildings can be reduced to 1 hour when the building is protected by an automatic sprinkler system.

There is an option to the control area concept available only to Group B higher education facilities. Chapter 38 deals specifically with the hazardous materials and hazardous operations in laboratories in higher education facilities. These facilities have the option of using control areas or laboratory suites. Laboratory suites in higher education facilities are discussed in detail in Chapter 15.

HAZARD IDENTIFICATION SIGNS

To ensure that emergency responders are aware of the presence of hazardous materials at a facility or building, the IFC requires the posting of hazard identification signs at certain locations. A hazard identification sign must comply with the format and classification criteria in NFPA 704, Standard System for the Identification of the Hazards of Materials for Emergency Response. This hazard identification sign is not allowed to be used for identifying the hazards of materials in transportation. An NFPA 704 hazard identification sign identifies the health, flammability and chemical instability of a particular hazardous material. An NFPA 704 diamond is divided into four colored fields (Figure 16-17). The color fields and hazards are

- Blue – health hazards – 9 o'clock position
- Red – flammable hazards – 12 o'clock position
- Yellow – chemical instability – 3 o'clock position
- White – special hazards – 6 o'clock position.

FIGURE 16-17 The NFPA 704 identification system provides a uniform format to communicate the health, flammability, chemical instability and special hazards to emergency responders.

The NFPA 704 system uses a numerical scale of 0 to 4 to indicate the relative hazards of hazardous materials. A value of 0 represents the least hazard versus a value of 4, which represents the greatest hazard. Table 16-5 summarizes the NFPA 704 hazard ratings based on the hazard categories. Additional guidance in IFC Appendix F correlates the IFC hazardous material classifications to the NFPA 704 ratings.

> **Code Essentials**
>
> Appendix F contains a cross reference between the IFC hazard category definitions and the appropriate designation of the NFPA 704 symbol.

TABLE 16-5 NFPA 704 hazard ratings by hazard categories

NFPA 704 hazard rating	Hazard categories		
	Health	Flammability	Instability
4	Materials that under emergency conditions can be lethal.	Materials that rapidly or completely vaporize at atmospheric pressure and normal ambient temperature or that are readily dispersed in air and burn readily.	Materials that in themselves are readily capable of detonation or explosive decomposition or explosive reaction at normal temperatures and pressures.
3	Materials that under emergency conditions can cause serious or permanent injury.	Liquids and solids (including finely divided suspended solids) that can be ignited under almost all ambient temperature conditions.	Materials that in themselves are capable of detonation or explosive decomposition or explosive reaction but that require a strong initiating source or must be heated under confinement before initiation.
2	Materials that under emergency conditions can cause temporary incapacitation or residual injury.	Materials that must be moderately heated or exposed to relatively high ambient temperatures before ignition can occur.	Materials that readily undergo violent chemical change at elevated temperatures and pressures.
1	Materials that under emergency conditions can cause significant irritation.	Materials that must be preheated before ignition can occur.	Materials that in themselves are normally stable but that can become unstable at elevated temperatures and pressures.
0	Materials that under emergency conditions would present no hazard beyond that of ordinary combustible materials.	Materials that will not burn under typical fire conditions, including intrinsically noncombustible materials such as concrete, stone, and sand.	Materials that in themselves are normally stable, even under fire conditions.
NFPA 704 special hazards			
NFPA 704 special hazard designations represent hazardous materials that may be water reactive, an oxidizer, corrosive, or a simple asphyxiation hazard. A water-reactive hazard is designated with a stricken-through W (W̶). Oxidizers can be represented with the letters "OX" while corrosives are designated as "COR" or "ALK" for alkali materials. Simple asphyxiants such as inert gases or inert cryogenic fluids are designated as "SA."			

The IFC adopts the NFPA 704 system because it provides a uniform means of estimating the hazards of materials at a particular facility or in a given area of a building. Because NFPA 704 provides a uniform method of identifying hazards to fire fighters, the system is easily implemented on local and regional levels. This is an important consideration, since hazardous materials response can involve regional deployment of emergency response personnel and equipment from multiple jurisdictions; uniformity in hazard identification is necessary to ensure hazards are accurately communicated to emergency responders. By using a standard format, many products are available in the marketplace that can assist businesses to easily comply with the hazard identification sign requirements. [Ref. 5003.5]

NFPA 704 hazard identification signs are required for above-ground storage tanks and stationary containers to identify the hazards of the stored material (Figure 16-18). Identification signs are required at specific facility entrances or locations required by the fire code official and in locations where the class and amount of hazardous materials exceed the permit quantities in Section 105.6. [Ref. 5003.5]

FIGURE 16-18 An NFPA 704 hazard identification sign is required for above-ground storage tanks storing hazardous materials.

SEPARATION OF INCOMPATIBLE MATERIALS

Accidental mixing of hazardous materials that are chemically incompatible is one cause of unauthorized discharges. It can result from an individual not understanding the hazards of mixing two chemicals together. It may result from personnel improperly modifying equipment or piping to accommodate a temporary situation where a stored energy source, such as a compressed gas, is required. Where materials have the potential to react in a manner that generates heat, fumes, gases or byproducts that are hazardous to life or property, they are considered incompatible. Recognize that certain manufacturing processes perform intentional chemistry, which is the processing of substances so an intended chemical reaction takes place. Intentional chemistry is classified as use by the IFC, and the fire code has specific requirements to ensure that the temperature, pressure, process flow rates and sequences of mixing are properly and safely conducted. [Ref. 202, 5005.1.11]

The provisions in Section 5003.9.8 address separating incompatible materials that may be accidentally mixed or come into contact in the event the primary container fails. The requirements are applicable to containers with a volume of more than 5 pounds or 0.5 gallon and all compressed gas containers, regardless of the cylinder or tank volume. [Ref. 5003.9.8]

A source of information for hazardous materials' chemical compatibility is the safety data sheet (SDS). The U.S. Department of Labor, Occupational Safety and Health Administration (OSHA) requires the SDS to include information concerning the material's reactivity, which can include classes of or specific materials that are incompatible with the chemical.

Certain classes of hazardous materials are considered to be some form of a reactive hazard, which raises the potential of an unauthorized discharge. Substances with the potential of spontaneously igniting, can form or are formulated as either an inorganic or organic peroxide, water-reactive material or self-reactive material. These particular materials must be closely evaluated for incompatibility with air, water and any contaminants that may not be readily apparent.

An electronic database in the ICC HMEX software allows one to evaluate the incompatibility hazards if two chemicals contact or are mixed together (Figure 16-19). The software does not include algorithms that contemplate mixing three or more hazardous materials. Such an operation should be treated as intentional chemistry.

> **You Should Know**
>
> The requirements in Chapter 50 are general requirements for hazardous materials. To properly regulate hazardous materials, Chapter 50 must be used in conjunction with the material-specific chapters 51 through 67.

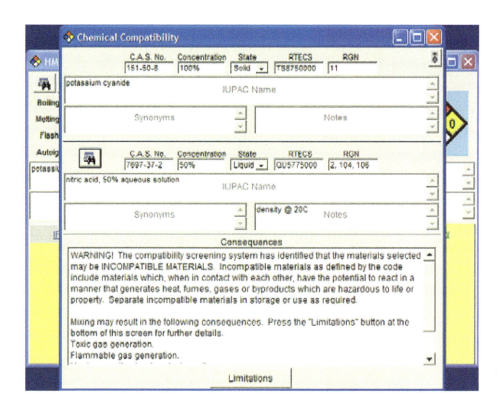

FIGURE 16-19 The HMEX software program allows users to determine if two materials mixed together are incompatible.

The IFC offers four methods of separating incompatible materials:
1. Separation can be accomplished by providing at least 20 feet of separation between the two incompatible materials.
2. Solids or liquids can be located in an approved hazardous material cabinet to separate them from the other material that is incompatible.
3. For gases, such as methane (a flammable gas) and oxygen (an oxidizer gas), one of the gas cylinders can be located into a gas cabinet or exhausted enclosure.
4. The IFC permits incompatible materials to be located adjacent to one another, provided they are separated by a noncombustible barrier that extends a minimum of 18 inches beyond and above each stored incompatible material (Figure 16-20). **[Ref. 5003.9.8]**

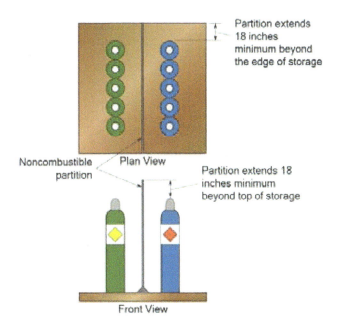

FIGURE 16-20 These incompatible hazardous materials are adequately separated by a noncombustible barrier.

CHAPTER 17
Compressed Gases

(Courtesy of Air Products and Chemicals, Allentown PA)

Gases are any substance that boils at atmospheric pressure and any temperature between absolute zero (-459.7°F) and up to about 80°F. Eleven chemical elements have boiling points within this temperature range: hydrogen, nitrogen, oxygen, fluorine, chlorine, helium, neon, argon, krypton, xenon and radon. Compressed gases are divided into two major groups, depending on their physical state in containers under certain pressures and temperatures and their range of boiling points: nonliquefied gases and liquefied gases. The *International Fire Code* (IFC) has a third category of compressed gas known as a dissolved gas; however, this group is limited to acetylene, which is a flammable gas. Acetylene is an unstable (reactive) material and must be stabilized for safe use. Acetylene is stored in a solution of acetone to stabilize the gas (Figure 17-1). **[Ref. 202]**

Nonliquefied or compressed gases do not liquefy at normal temperatures and under storage pressure that ranges up to about 6,000 pounds per square inch gauge (PSIG). Compressed gases can become a liquid if cooled below their boiling points. When gases are liquefied with boiling points below -130°F, they are referred to as cryogenic fluids and are regulated by IFC Chapter 55. Examples of compressed gases that can be converted to cryogenic fluids include nitrogen, argon, hydrogen and oxygen.

Liquefied compressed gases become liquids at ordinary temperatures and pressures from 25 to 2,500 PSIG. Liquefied gases are elements or compounds that have boiling points relatively near atmospheric temperatures, ranging from approximately -130°F to 25–30°F. Liquefied compressed gases would become solid at the low temperatures used for storing cryogenic fluids. Liquefied gases are packaged and transported under rules that limit the maximum amount that can be packaged into a container to allow space for liquid expansion when ambient temperatures rise. Examples of liquefied compressed gases include anhydrous ammonia, propane and carbon dioxide.

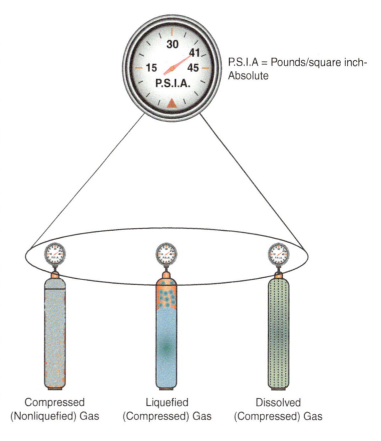

FIGURE 17-1 Compressed, liquefied and dissolved gases

IFC Chapter 53 defines and sets forth requirements for the storage, use and handling of compressed gases in gas containers, cylinders, tanks and systems. The requirements in Chapter 53 are applicable to the following hazard classes of gases:

- Asphyxiant
- Corrosive
- Flammable, including liquefied petroleum gases
- Highly toxic
- Inert
- Irritant
- Oxidizing
- Pyrophoric
- Radioactive
- Unstable (reactive)
- Toxic.

Proper application and enforcement of the Chapter 53 requirements are accomplished through the application of the storage and handling requirements in Chapter 53, the material-specific requirements in IFC Chapters 51 through 67 and the general hazardous material requirements in Sections 5001 and 5003. **[Ref. 5301.1]**

The intent of the Chapter 53 requirements is to safely control the release of the potential energy stored in a compressed gas container. Gases are commonly stored at pressures of 25 to over 6,000 PSIG. The accidental release of a compressed gas and its stored potential energy can be instantaneous. This rapid, violent energy release can easily propel a 125-pound gas cylinder through the air as a missile; the projectile cylinder and debris can injure or kill individuals.

The requirements in Chapter 53 ensure the cylinder is correctly designed and constructed to safely contain the volume of filled gas at the required delivery pressure, the cylinder valve is constructed with a unique fitting to prevent it from being connected to the wrong type of pressure regulator and the compressed gas source and system to which it is connected are operated within predetermined safe temperature, pressure and flow limits (Figure 17-2). The IFC requires protection of the cylinder and its connected components from mechanical impact, fire exposure and other conditions or external exposures that could compromise the pressure containment system. The IFC prohibits the use of compressed gas containers and systems that are leaking, fire damaged or improperly assembled or modified.

FIGURE 17-2 A properly engineered and constructed compressed gas storage and distribution system

CYLINDERS, CONTAINERS AND TANKS

In the context of IFC Chapter 53, a compressed gas container is a pressure containment vessel. The container must be engineered to safely store and dispense the gas based on its physical or health hazards and its physical properties. Compressed gas containers are constructed and maintained in accordance with approved standards and, unless they are disposable, are subject to periodic testing and examination. Compressed gas cylinders are to be designed and constructed in accordance with the requirements of the U.S. Department of Transportation (DOT) or the American Society of Mechanical Engineers (ASME)'s *Boiler and Pressure Vessel Code* requirements. **[Ref. 5303.2]**

Regardless if it is a cylinder, container or tank, all compressed gas containers have common and standard features:

- The size of a cylinder or container is limited. If it is a tank, like a stationary pressure vessel, its size is generally not limited.
- The volume of gas that can be stored is directly proportional to the density of the gas molecules. The lighter the gas molecules, the greater the volume of gas that can be stored when compared to heavier gas molecules stored in the same volume.
- Cylinders, containers and tanks always have a circular cross-

sectional area; they are not constructed using other geometric shapes because of the imposed forces in the corners of rectangular or square-shaped containers.
- Proper storage and use of compressed gas containers is dependent on their orientation. Cylinders are designed for either a vertical or horizontal orientation; specially designed cylinders can be stored and used in either orientation (Figure 17-3).
- With the exception of a limited number of physical or health hazard materials, all compressed gas containers are equipped with a pressure relief device. The pressure relief device protects the cylinder from an explosion resulting from excessive heating of its contents.
- Valve fittings are equipped with unique, gas-specific connections to prevent the connection of the gas cylinder to the wrong pipe or distribution system.
- Cylinders require markings that identify the standard to which they are constructed, that they are qualified for service, its contents and the basic hazards of the stored gas.

FIGURE 17-3 A stationary pressure vessel designed to be installed vertically

PRESSURE RELIEF DEVICES

With few exceptions, all compressed gas containers are equipped with a pressure relief device (PRD). A PRD is a safety device that protects a gas container from catastrophic failure resulting from it being overpressurized. Overpressurizing a compressed gas cylinder or tank can occur if the container is directly involved in or exposed to a fire, or if it is overfilled. PRDs provide a safe, reliable method of depressurizing gas containers so they do not explode or become a projectile. [Ref. 5303.3.1]

PRDs must be designed and installed as in accordance with one of the following design standards:
- American Society of Mechanical Engineers' *Boiler and Pressure Vessel Code*
- Compressed Gas Association S-1.1, Pressure Relief Device Standards—Part 1—Cylinders for Compressed Gases
- Compressed Gas Association S-1.2, Pressure Relief Device Standards—Part 2—Cargo and Portable Tanks for Compressed Gases
- Compressed Gas Association S-1.3, Pressure Relief Device Standards—Part 3—Stationary Storage Containers for Compressed Gases

There are two general forms of PRDs: reusable devices or sacrificial devices. The type of PRD selected is based on the size of the gas container and the characteristics and hazards of the stored gas. The most common reusable PRD is a spring-loaded safety relief valve (Figure 17-4). Sacrificial PRDs are either a rupture disk or a fusible plug.

Code Essentials

With few exceptions, all cylinders and gas containers require a means of pressure relief. Pressure relief devices are either reusable or sacrificial. Spring-loaded pressure relief valves are one type of a reusable pressure relief device. Burst disks and fusible plugs are sacrificial relief devices.

FIGURE 17-4 An internal spring-loaded pressure relief valve. The valve is installed inside the gas container.

FIGURE 17-5 External spring-loaded pressure relief valves

FIGURE 17-6 Front view of a 1-ton capacity compressed gas container. The cylinder is equipped with three fusible plugs located at the 9 o'clock, 1 o'clock and 5 o'clock positions. Each fusible plug is constructed by machining a hole inside a pipe plug and filling the hole with a low melt-point metal.

FIGURE 17-7 A burst disc installed on top of a cryogenic nitrogen container. If the container becomes overpressurized, the disc will burst open to safely release the cryogenic fluid.

Spring-loaded PRDs utilize a valve that is maintained closed by a tensioned spring. The valve remains closed during normal cylinder use and filling. If the pressure inside the container exceeds a predetermined set point, the spring is compressed; this action opens a valve releasing some of the gas and lowering the internal pressure. When the container's internal pressure is reduced below the PRD set point, the valve closes, stopping the release of gas. The PRD will continue to cycle open and close until enough gas is discharged and the internal pressure is reduced (Figure 17–5).

When a sacrificial PRD operates, it cannot reclose. Operation of a sacrificial PRD expels the entire contents of the gas container. Sacrificial PRDs operate when the device's pressure or temperature limits are exceeded. Heat-activated PRDs are manufactured as fusible relief valves, known as fusible plugs. Fusible plugs operate when they are subjected to direct or indirect heat from a fire. Fusible plugs are set to operate at temperatures between 130–350°F and are assembled using metals with extremely reliable and predictable melting temperatures (Figure 17-6).

Sacrificial pressure sensitive PRDs are known as burst discs or rupture discs. A burst disc is an engineered device that uses a metal or graphite disc that is weakened by scoring (Figure 17-7). Scoring is a mechanical process where the metal is partially cut to cause a predictable failure at the opening pressure of the burst disc. The disc's size, geometry and material of construction governs its opening or "burst" pressure. When a burst disc operates, it opens like a flower and the stored gas is released to the atmosphere (Figure 17-8).

Regardless of the type of PRD used, it must be properly installed and maintained. Sizing of PRDs is governed by the specification to which the container was fabricated and is based on the flow characteristics and the hazards of the stored gas. After the gas container empties, the sacrificial PRD is replaced. [Ref. 5303.3.3]

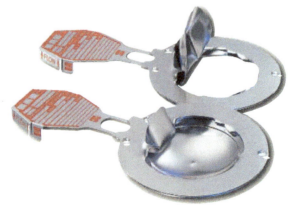

FIGURE 17-8 Unopened and opened burst discs. The top burst disc has operated. *(Courtesy of BS&B Safety Systems, Tulsa, OK)*

MARKINGS

Stationary and portable gas containers, cylinders and tanks require markings. Markings demonstrate that cylinders or tanks are qualified to safely store and contain compressed gases. In the case of portable compressed gas cylinders, markings are used to identify that containers have been subjected to periodic pressure tests and inspections as prescribed by the U.S. Department of Transportation (DOT) regulations and laws. Markings communicate the hazards of the stored gas to consumers and emergency responders. When compressed gas containers are connected to a gas distribution system, markings are also required on the piping network so maintenance personnel are aware of the pipe contents and to assist fire fighters who may be called in the event of a system leak. [Ref. 5303.4]

Stationary containers, like pressure vessels, require markings to identify their NFPA 704 hazard rating as well as the actual contents of the container (Figure 17-9). Signs are also required to indicate that open flames or other ignition sources are not permitted near containers storing flammable gases. When the containers are located within a room or enclosure, the room or cabinet must also be marked with the words "COMPRESSED GAS." [Ref. 5003.5.1, 5303.4.1]

Portable containers, cylinders and tanks have various markings that identify the standard that governed their construction and design pressure. Other markings indicate that the vessel is qualified for the compressed gas it contains. These requirements are based on the hazardous material regulations in 49 CFR Parts 100–185 promulgated by the U.S. DOT. The markings will indicate the DOT specification that the cylinder was constructed to, its type and material of construction, the cylinder's service pressure measured in PSIG, the manufacturer's mark and the cylinder's serial number. Additional markings are required to indicate the month and year the cylinder was manufactured and cylinder testing requirements (Figure 17-10).

A common consumer misconception is that the color of a cylinder indicates its hazards. This assumption is not correct. A cylinder's color has no bearing on or correlation with what gas it may store. The only means of identifying the contents of a cylinder is by labeling or marking the cylinder. Cylinders require markings in accordance with Compressed Gas Association (CGA) Standard C-7, Guide to the Preparation of Precautionary Labeling

> **You Should Know**
>
> The type of markings required on compressed gas containers depends on whether the container is a portable cylinder or stationary tank. Piping connected to compressed gas sources also requires markings to indicate the pipe contents and direction of flow.

FIGURE 17-9 Hazard identification markings on a liquid (cryogenic) nitrogen container

U.S. DOT Cylinder Marking information

1. **Cylinder Specifications**
   - DOT — U.S. Dept. of Transportation. (Regulatory body that governs the use of cylinders.)
   - 3AA — Type and material of construction.
   - 2015 — Service pressure in pounds per square inch gauge.
2. A-13016 — Serial number
3. SRL — Identifying symbol, registered with DOT.
4. **Manufacturing Data**
   - 4-76 — Date of manufacture and original test date.
   - H — Inspectors official mark.
   - + — Cylinder qualifies for 10% overfill.
   - * — Cylinder qualifies for 10 year retest interval.

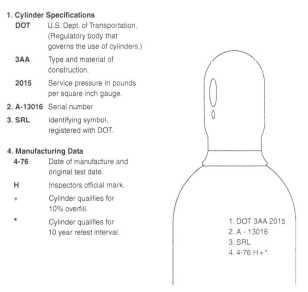

1. DOT 3AA 2015
2. A - 13016
3. SRL
4. 4-76 H + *

FIGURE 17-10 Cylinder markings required by U.S. DOT

and Marking of Compressed Gas Containers. These labels are located on the shoulder or wall of the cylinder. CGA Standard C-7 requires the marking contain the name of the hazardous material and its DOT and UN hazardous material identification label and hazard division identification number, which is used in the DOT *Hazardous Materials Emergency Response Guidebook* (Figure 17-11). If the compressed gas is an inhalation hazard, warning statements about this hazard are required, and its reportable quantity (RQ), if such a value is assigned by the U.S. DOT. An RQ is a value assigned by the U.S. DOT, and if the amount of hazardous material released exceeds the limit, it must be reported to the United States Coast Guard National Response Center. A release of hazardous materials in amounts greater than the RQ triggers a response by the federal government. Depending on the gas and amount of the release, the incident may result in a response by one or more federal agencies. [Ref. 5303.4.2]

FIGURE 17-11 A CGA C-7-compliant cylinder shoulder label for acetylene, a flammable gas *(Courtesy of Label Solutions, Inc.)*

FIGURE 17-12 Pipe labels indicate the direction of flow and the contents (anhydrous ammonia) in this mechanical refrigeration system.

Piping systems carrying compressed gases to a point of use are required to be marked with the name of the product being conveyed and the direction of flow. The markings must comply with ASME A13.1, Scheme for the Identification of Piping Systems. Pipe labels are required at each valve, at penetrations of a wall, floor or ceiling and at a maximum spacing interval of 20 feet along the entire length of the pipe run (Figure 17-12). [Ref. 5303.4.3]

SECURITY

To limit the likelihood of damage to compressed gas containers, the IFC requires cylinders be properly secured and protected from physical impact. The fire code prescribes three types of protection for compressed gas containers: security to safeguard the gas containers, physical protection from mechanical impact and a method of ensuring gas cylinders are protected from being knocked over or falling (Figure 17-13). [Ref. 5303.5]

Physical protection is required when compressed gas cylinders or containers could be subjected to mechanical impact. Impact can be in the form of an object being accidentally dropped, such as the movement of materials by a crane over the compressed gas cylinders or physically striking the cylinders with material-handling equipment. If containers or stationary tanks are located outdoors in an area subject to impact by vehicles, physical protection consisting of guard posts or similar barriers is required. IFC Section 312.1 provides criteria on methods of protecting these cylinders or tanks from vehicular impact (Figure 17-14). [Ref. 5303.5.2]

Cylinders stored and used in either a vertical or horizontal orientation require protection from falling as a result of being accidentally impacted, equipment vibration or seismic ground motion. The IFC recognizes the following methods to protect cylinders from falling:

- Securing cylinders or containers to a fixed object with one or more methods of restraint.
- Securing the cylinders or containers on a cart or other device designed for the movement of compressed gas containers or cylinders.
- Nesting of cylinders, which is a method of securing flat-bottomed compressed gas cylinders upright in a tight mass using a contiguous three-point contact system whereby all cylinders within a group have a minimum of three points of contact with other cylinders, walls or bracing (Figure 17-15).
- Securing cylinders in a gas cabinet, storage rack or frame (Figure 17-16). [Ref. 3003.5.3]

FIGURE 17-13 This cylinder exchange storage cabinet is designed to protect the cylinders from tampering and is constructed of expanded metal to provide ventilation and so hose streams can apply water to cool cylinders that may be involved in a fire. Guard posts protect the LP-gas cylinders from vehicular impact.

FIGURE 17-14 Vehicle impact protection is provided for this stationary pressure vessel storing carbon dioxide.

Code Essentials

Physical protection is required for all compressed gas cylinders and tanks. This includes protection from falling, vehicular impact, being dislodged during a seismic event or being hit with water stream during fire-fighting operations.

254 Chapter 17 Compressed Gases

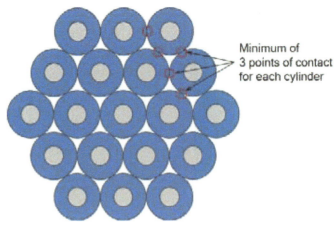

FIGURE 17-15 Properly nested cylinder storage requires that all of the cylinders have at least three points of contact.

FIGURE 17-16 A cylinder storage rack

VALVE PROTECTION

To protect a cylinder from becoming a flying projectile, the IFC requires cylinder valves be protected when they are moved or are in storage. Valve protection is accomplished by either shielding the valve or installing a protective cap. Valve shields are installed directly on the compressed gas container and are constructed with an opening to facilitate installation or removal of pressure regulators or piping. If a cylinder is not constructed with a valve shield, it is equipped with a protective valve cap (Figure 17-17). [Ref. 5303.6.1]

SEPARATION FROM HAZARDOUS CONDITIONS

The IFC sets forth requirements that are intended to protect the compressed gas system from exposure to hazardous conditions. This includes incompatible hazardous materials, excessive heating of the cylinders, physical damage and ignition sources.

As previously indicated in Chapter 16, IFC Section 5003.9.8 has requirements for the separation of incompatible hazardous materials, which include compressed and liquefied compressed gases. For certain hazard classes of gases, the code specifies additional requirements that further limit the potential for incompatible hazardous material storage. The IFC limits the cylinder volume of highly toxic and toxic gases in assembly, business, educational, institutional, mercantile and residential occupancies to 20 cubic feet and requires they be stored in a gas cabinet or exhausted enclosure in those occupancies. Placing these gases in gas cabinets or exhausted enclosures

FIGURE 17-17 A shield protects the valve on a 20-pound liquefied petroleum gas container.

satisfies the requirement for separation of incompatible materials. Pyrophoric gases are separated from incompatible hazardous materials by 1-hour fire barriers unless they are stored in approved hazardous material storage cabinets, gas cabinets or exhausted enclosures. [Ref. 5303.7.1, 6404.1.4]

Cylinders, containers and tanks must be protected from excessive temperatures unless they are designed to operate in such conditions. Cylinders must be stored in an area where surface temperatures do not exceed 125°F. The intent of the requirements is to reduce the potential of the gas pressure increasing to a level that could cause the container's PRD to operate. In areas where extreme temperatures occur, the IFC requires that a cover or shade structure be provided to protect the cylinders. [Ref. 5303.7.4, 5303.14]

Certain gas cylinders or their connected piping may be equipped with a device or system to heat the container. Heating the cylinder increases the pressure of the gas, which allows for it to be safely distributed. Cylinder heaters are equipped with a fail-safe control to ensure the temperature does not exceed 125°F and can only be used by properly trained personnel. They are installed in accordance with the requirements in the *International Mechanical Code* (IMC) and the *National Electrical Code* (Figure 17-18). Cylinders are not to be exposed to temperatures above 125°F. This is the surface temperature of the cylinder or tank. Keep in mind that the air temperature does not need to reach 125°F for the cylinder to be exposed to these high temperatures. Storage inside or a shade structure is often required to limit exposure to this temperature. [Ref. 5303.7.4, 5303.7.6]

FIGURE 17-18 A jacket cylinder heater

A common method of securing compressed gas cylinders, containers and tanks from falling is nesting. When nesting is used on an elevated platform or near the edge of a building such as a loading dock, the gas containers must be adequately separated from the edge of the platform or shaft opening to limit the potential for damage if the cylinder or container falls (Figure 17-19). The minimum separation distance is one-half of the height of the cylinder. [Ref. 5303.7.3]

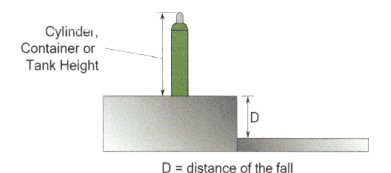

FIGURE 17-19 Cylinders must be protected or moved when near the edge of a platform or loading dock.

EXHAUSTED ENCLOSURES AND GAS CABINETS

Exhausted enclosures and gas cabinets are noncombustible appliances designed to capture and safely exhaust a leak from a compressed gas container or its connected piping systems. They are one of several methods permitted by Tables 5003.1.1(1) and 5003.1.1(2) that allow a 100 percent increase in the maximum allowable quantity per control area (MAQ) for most gases. Gas cabinets and exhausted enclosures provide a means of separating incompatible hazardous materials and provide protection to comply with the requirements in Section 5303.7 for separating cylinders and containers from hazardous conditions.

The difference between an exhausted enclosure and a gas cabinet is the number of sides that are open. An exhausted enclosure is an appliance that is constructed to form a top, sides and back. The front, or a portion of the front, is open, which allows for access to the cylinders. The exhausted enclosure is equipped with mechanical ventilation to capture and exhaust gases or fumes so they do not escape through the open front. Examples of exhausted enclosures include laboratory hoods or exhausted fume hoods (Figure 17-20). A room or area provided with a hazardous exhaust mechanical ventilation system is not an exhausted enclosure. [Ref. 5003.8.5, 5303.7.9]

FIGURE 17-20 A laboratory hood is a type of exhausted enclosure.

A gas cabinet is a noncombustible enclosure that provides an isolated environment for compressed gas cylinders in storage or use (Figure 17-21). Gas cabinets are constructed with doors and windows that allow access to operate pressure regulators or for exchanging cylinders. They are constructed of at least 12-gage steel with self-closing access ports or windows and doors and must be equipped with a mechanical ventilation system that maintains a negative pressure in relation to the surrounding room or area. Not more than three compressed gas cylinders are permitted in a single gas cabinet. [Ref. 5003.8.6, 5303.7.10]

FIGURE 17-21 A gas cabinet

LEAKS, DAMAGE OR CORROSION

A basic rule of any hazardous material system, regardless of the chemical's physical state, is that leaks rarely if ever become smaller—they generally become bigger. A leak in a compressed gas system presents serious fire safety and health implications: the atmosphere in the room or area can become toxic, oxygen deficient or can increase the possibility of a fire or explosion. Because of these risks, the IFC establishes a zero-tolerance requirement that prohibits the use of leaking, damaged or corroded gas containers. If these conditions are identified, the cylinder must be removed from service and repaired to a serviceable condition or otherwise be replaced. [Ref. 5303.12]

The term *corrosion* relates directly to the reduction of the thickness of carbon or stainless steel, such as in the wall or head of a compressed gas container, cylinder or storage tank. Any unprotected carbon steel exposed to humidity in the air eventually will have external corrosion. Rust should not be the determining factor for whether or not the equipment should be replaced. Verification and confirmation of internal corrosion will require the measurement of the metal's thickness and a hydrostatic pressure test to determine that the container's pressure containment ability has not been reduced or compromised.

A compressed gas container that has been exposed to a fire must be removed from service. Fire exposure of a compressed gas container can cause sacrificial pressure relief devices to operate, which must be replaced. The heat of a fire can change the metallurgy of the container, which could reduce its strength or resistance to mechanical impact. Fire-damaged containers, cylinders or tanks should be removed by approved qualified personnel who have experience dealing with damaged compressed gas containers (Figure 17-22). [Ref. 5303.11]

FIGURE 17-22 An LP-gas storage tank damaged by a wildland fire. This tank was removed from service, subjected to a hydrostatic pressure test and all of its valves and fittings were replaced before it was returned to service.

LP-GAS CYLINDER EXCHANGE PROGRAM

A common sight at many retail facilities is the LP-gas cylinder exchange cabinets. The cylinder exchange program, where the consumer returns an empty cylinder and takes a full cylinder, has been around for decades. However, in 1998 NFPA 58, *Liquefied Petroleum Gas Code*, was revised to prohibit the refilling of cylinders from 4 pounds to 40 pounds not equipped with an overfill prevention device (OPD) (Figure 17-23). This led to the current exchange program that results in the following scenario:

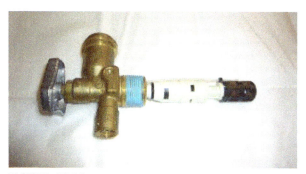

FIGURE 17-23 LP-gas cylinder overfill prevention device (OPD) *(Courtesy of National Propane Gas Association, Washington, DC)*

- A consumer takes an empty 20-pound LP-gas cylinder to a cylinder exchange location, drops off the empty and returns home with a full cylinder.
- The LP-gas refill company takes the empty (returned) cylinder and evaluates it.
- The cylinder must pass a visual inspection (the company looks for

> **You Should Know**
>
> Chapter 53 requirements for compressed systems address the protection of the cylinder and ensure it can safely contain its stored energy. The material-specific requirements in IFC Chapters 51 through 67 must be applied in conjunction with the Chapter 53 requirements for compressed gases. In general, all cylinders require
>
> - Construction in accordance with approved standards
> - A pressure relief device
> - Markings to indicate the container, cylinder or tank is constructed in accordance with approved standards and the hazards of the stored compressed gas
> - Security and physical protection of the container and container valve
> - Separation from hazardous conditions
> - Proper maintenance. If a compressed gas container is leaking, damaged or has been exposed to a fire, it must be removed from service and be repaired to a serviceable condition or be replaced. •

signs of damage).
- The cylinder is hydrostatically tested, depending on the date of manufacture or the date of the previous test.
- If the cylinder passes the visual and hydrostatic tests, the cylinder will be retrofit with an OPD.
- Acceptable cylinders will be refilled and placed back into a cylinder exchange cabinet (Figure 17-24).

FIGURE 17-24 An LP-gas cylinder exchange station

According to the National Propane Gas Association, this process has resulted in nearly one million cylinders being removed from service that did not pass the inspection and testing criteria.

An OPD is designed to keep the LP-gas cylinder from being overfilled. The contents of the LP-gas cylinder consist of a combination of gas and liquid. The LP-gas cylinder must be provided with a vapor space to allow changes in pressure as the surrounding temperature changes. An OPD is calibrated to stop the flow into the cylinder when the maximum fill level is reached. The OPD must be listed to ANSI/UL 2227, Standard for Overfilling Prevention Devices. **[Ref. 6106.2]**

As the cylinder exchange program has expanded, the cylinder exchange stations have become automated. The customer can complete the exchange and pay with a credit card at the kiosk. The code requires that the exchange components and equipment be inspected daily. The cylinder cabinet must be designed so that the returned cylinder can only be inserted in an upright position. This ensures that the pressure relief device is in the vapor space rather than in the residual liquid that may be present in the cylinder. All electrical equipment within 5 feet of the cylinder storage cabinet must be designed as Class I, Division 2 (Figure 17-25). **[Ref. 6109.15.1]**

FIGURE 17-25 An automated LP-gas cylinder exchange station

LIQUID CARBON DIOXIDE FOR BEVERAGE DISPENSING

A trend in restaurants and other businesses serving carbonated beverages is to store carbon dioxide as a liquid rather than as a gas. The advantage to the business owner is that space that was occupied by multiple carbon dioxide cylinders is now available for other storage because the multiple cylinders were replaced with one insulated dewar storing liquid carbon dioxide (Figure 17-26).

The business owner no longer has to exchange carbon dioxide cylinders to keep up with use. Carbon dioxide is delivered to the business in liquid form and the dewar is refilled via an exterior connection (Figure 17-27).

> **You Should Know**
>
> A dewar (pronounced do-er) is essentially a large thermos. It is a double-walled vessel with the interstitial space under vacuum. It is designed to hold extremely low-temperature liquids and is vented to the atmosphere.

FIGURE 17-26 This dewar stores carbon dioxide as a liquid and is connected to the beverage dispensing system.

FIGURE 17-27 This remote connection is located outdoors. The photo on the left is the remote fill connection with the cover closed. The cover has been opened to reveal the carbon dioxide hose connection on the right.

> **Code Essentials**
>
> Carbon dioxide as a refrigerated liquid is <u>not</u> considered a cryogenic fluid. Carbon dioxide has a boiling point of -109°F.
> To be considered a cryogenic fluid, the material's boiling point must be lower than -130°F.

Carbon dioxide is odorless and colorless. Even though it is stored as a liquid, a release will rapidly convert to gas. The gas is heavier than air and is an asphyxiant. Many of these liquid carbon dioxide systems for beverage dispensing are installed in small rooms and basements. This placement creates a hazard for employees and customers when an accidental release occurs.

Systems with more than 100 pounds of liquid carbon dioxide for beverage dispensing require an operational permit. As part of the permit approval, dewars and systems installed indoors are required to provide mechanical ventilation which operates continuously or a gas detection system. The gas detection system requires two alarm thresholds:

1. Low-level alarm at 5,000 ppm: permissible exposure limit (PEL)
2. High-level alarm at 30,000 ppm: threshold limit value – short term exposure limit (TLV-STEL). **[Ref. 5307.3.1, 5307.3.2]**

Activation of the low-level alarm results in an audible and visible alarm operating in a normally attended location. The high-level alarm activates an audible and visible alarm in the room or area where the liquid carbon dioxide system is installed.

CHAPTER 18
Flammable and Combustible Liquids

Of all the physical and health hazard materials available to private and industrial consumers, flammable and combustible liquids are the most abundant and easily accessible. Flammable and combustible liquids are used as transportation fuels, lubricants, distilled spirits, chemical feedstock in the manufacturing of plastics, pharmaceutical, semiconductors and a host of other beneficial uses.

Flammable and combustible liquids are the most regulated hazardous material in the *International Fire Code* (IFC). Requirements for storage, use, handling and dispensing are found in Chapter 57, along with other chapters regulating specific operations. The IFC references portions of NFPA 30, *Flammable and Combustible Liquids Code*, and NFPA 30A, *Code for Motor Fuel Dispensing Stations and Repair Garages*, for specific regulations. Because flammable and combustible liquids have such a wide range of uses, other IFC chapters also have requirements when they are used for specific purposes. Other provisions that regulate specific flammable and combustible liquid uses are found in

- IFC Chapter 6—Fuel-Fired Appliances and Commercial Kitchen Cooking-Oil Storage
- IFC Chapter 20—Aviation Facilities
- IFC Chapter 21—Dry Cleaning
- IFC Chapter 23—Motor Fuel-Dispensing Facilities and Repair Garages
- IFC Chapter 24—Application of Flammable Finishes
- IFC Chapter 27—Semiconductor Fabrication Facilities
- IFC Chapter 29—Manufacture of Organic Coatings
- IFC Chapter 33—Fire Safety During Construction and Demolition
- IFC Chapter 38—Higher Education Laboratories
- IFC Chapter 39—Processing and Extraction Facilities.

CLASSIFICATION OF LIQUIDS

Liquids regulated by IFC Chapter 57 are divided into two groups: flammable and combustible. A liquid is assigned to either group based on its boiling point and flash point temperatures. A liquid's boiling point is "the temperature at which the vapor pressure of a liquid equals the absolute pressure of 14.7 pounds per square inch absolute (PSIA)." The boiling point temperature is a measurement of the volatility of the liquid and is one method of measuring how fast or slow a liquid evaporates. The lower a liquid's boiling point temperature, the more rapidly it will produce vapors. This is an important consideration because flammable and combustible liquids themselves do not burn—it is their vapors that ignite and burn. [Ref. 202]

Flash point temperature is the minimum temperature at which a liquid will give off sufficient vapors to form an ignitable mixture with air near its surface or in the container but yet will not continue to burn after the vapors flash across the surface. It is the basis for classifying a liquid as either flammable or combustible. Flash point is one measure of the tendency of the liquid test specimen to form a flammable mixture with air under controlled laboratory conditions. It is only one of a number of properties that should be considered in assessing the overall flammability hazard of a liquid.

Flash point temperature is determined by tests performed in a specialized instrument. In a flash point test apparatus, a small amount of liquid is incrementally heated to a specific temperature. As the liquid temperature is raised, a pilot flame is swept directly over the test sample. If the pilot flame ignites the vapor, a flame flashes across the liquid surface. This event does not cause the vapor to continue to sustain burning and is the flash point temperature of the liquid. [Ref. 202]

The IFC requires closed cup flash point testing be performed in accordance with one of four ASTM standards. Closed cup tests

> **Code Essentials**
>
> Flammable and combustible liquids are categorized into specific hazard classifications based on their
> - flash point temperature and
> - boiling point temperature.

are required by the IFC and NFPA 30 because they provide a higher degree of reproducibility. The results of closed cup flash point tests must be reproducible within ±6°F versus ±18°F for open cup flash point tests (Figure 18-1). [Ref. 202]

Flammable and combustible liquids are further divided into different individual classes that differentiate their flash point and boiling point temperatures. These additional categories are used to assign the relative fire hazard of liquids and serve as the basis for determining the maximum allowable quantity of liquid allowed in a control area (see Chapter 16 of this text for additional information on maximum allowable quantity). Table 18-1 presents the classification for the different classes of flammable and combustible liquids.

As previously indicated, liquids do not burn—their vapors burn. For ignition to occur, the volume of vapor must be within a given

FIGURE 18-1 Cleveland Open Cup Flash Point test apparatus *(Courtesy of AMETEK PetroLab, Albany, NY)*

TABLE 18-1 Classification of flammable and combustible liquids

Classification	Flash point temperature	Boiling point temperature	Example liquids by class
Flammable liquid			
Class IA	< 73°F	< 100°F	N-Pentane; Ethyl ether
Class IB	< 73°F	≥ 100°F	Gasoline; Isopropyl alcohol
Class IC	≥ 73 and < 100°F	Not applicable	Methyl ethyl ketone
Combustible liquid			
Class II	≥ 100°F and < 140°F	Not applicable	No. 2 Diesel fuel; Kerosene
Class IIIA	≥ 140°F and < 200°F	Not applicable	Turpentine; No. 3 fuel oil
Class IIIB	≥ 200°F	Not applicable	Motor oil; Cooking oil; No.6 (Bunker) fuel oil

flammability range in air. All flammable and combustible liquids have a flammable range in air and these ranges are based on a material's lower and upper flammable limits. The lower flammable limit (LFL) is the lowest concentration at which ignition of the vapor-air mixture can occur. The upper flammable limit (UFL) is the highest concentration at which ignition of the vapor-air mixture can occur. Ignition cannot occur when the flammable vapor-air mixture is less than LFL or greater than UFL. A liquid's flammable range is based on its flash point and boiling point temperature, vapor pressure, molecular weight and the arrangement of its molecules. Mixing of liquids will influence the flammable range as well as its flash and boiling point temperatures. The flammable range of any liquid or gas is determined by testing in accordance with one of several ASTM standards. The flammable range of liquids and gases will change if

Code Essentials

All flammable and combustible liquids exhibit a flammable range that is based on its lower flammable limit (LFL) and upper flammable limit (UFL). ●

the atmosphere is enriched with oxygen or if it is diluted with an inert gas or inert liquid such as water (Figure 18-2).

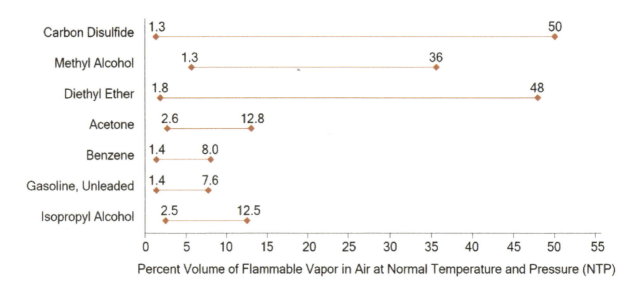

FIGURE 18-2 The flammable range of seven liquids

> ### QUESTIONS
> Carbon disulfide is used for the manufacturing of rayon and cellophane. The liquid has a boiling point temperature of 115°F and closed cup flash point temperature of 29°F.
>
> A. Is carbon disulfide a flammable or combustible liquid?
> B. What is its classification?
>
> ### ANSWERS
> A. Flammable liquid
> B. Class IB flammable liquid

CONTAINERS, PORTABLE TANKS AND STATIONARY TANKS

Flammable and combustible liquids are packaged and stored in containers, portable tanks or stationary tanks. These terms are not interchangeable, and understanding the difference between each term is an important part of code enforcement. The materials of construction, volume and any safety features are dependent on the hazards of the liquid, specific consumer requirements such as product purity and if the vessel is designed for transportation. If a container or portable tank can be used for transportation, it must be designed and constructed in accordance with U.S. Department of Transportation (DOT) requirements. Stationary tanks are regulated by the requirements in the IFC and specific sections of NFPA 30, which are referenced in IFC Chapter 57.

The IFC limits the volume of containers and portable tanks, and NFPA 30 has specific requirements based on the classification of the liquid and the stored volume. A container is a vessel of 60 gallons or less in capacity used for transporting or storing hazardous materials. A drum is a vessel of 119 gallons or less in capacity used for transporting or storing hazardous materials. When the IFC refers to containers, the reference includes drums only if they are 60 gallons or less in capacity. Above 60 gallons, the drum is treated as a portable tank. Flammable and combustible liquid containers and drums are constructed of glass, plastic, fiberboard or steel in accordance with the requirements in NFPA 30 and U.S. DOT. Pipes, piping systems and engine fuel tanks are not considered by the IFC to be containers. The IFC refers to NFPA 30 for construction and maximum size of containers and portable tanks. Table 18-2 indicates the maximum size of containers and portable tanks based on their construction and the hazard of the contents. NFPA 30 limits the volume of glass and plastic containers and drums storing Class IA, IB, IC and II liquids because these containers can rapidly fail if they are subject to fire exposure (Figures 18-3 and 18-4). **[Ref. 5003.2.1, 5704.3.1]**

FIGURE 18-3 The allowed volume of glass containers is restricted by the IFC based on the classification of the stored liquid.

FIGURE 18-4 The allowed volume of metal containers can be greater since the level of fire safety is increased and the potential for a spill is decreased compared to glass containers.

TABLE 18-2 Maximum size of container, portable tank and intermediate bulk container

Container type	Flammable liquids			Combustible liquids	
	Class IA	Class IB	Class IC	Class II	Class IIIA
Glass	1 pint	1 quart	1.3 gallons	1.3 gallons	5.3 gallons
Metal (other than drums) or approved plastic	1.3 gallons	5.3 gallons	5.3 gallons	5.3 gallons	5.3 gallons
Safety cans	2.6 gallons	5.3 gallons	5.3 gallons	5.3 gallons	5.3 gallons
Metal drum	119 gallons	119 gallons	119 gallons	119 gallons	119 gallons
Approved metal portable tanks and IBCs	793 gallons	793 gallons	793 gallons	793 gallons	793 gallons
Rigid plastic IBCs and composite IBCs with rigid inner receptacle	NP	NP	NP	793 gallons	793 gallons
Composite IBCs with flexible inner receptacle	NP	NP	NP	NP	NP
Non-bulk Bag-in-Box	NP	NP	NP	NP	NP
Polyethylene	1.3 gallons	5.3 gallons	5.3 gallons	119 gallons	119 gallons
Fiber drum	NP	NP	NP	119 gallons	119 gallons

Code Essentials

- A container has a volume of 60 gallons or less.
- A portable tank has a volume of 60 to 793 gallons.
- Containers and portable tanks are designed for transportation.
- A stationary tank is designed for permanent above ground, underground and indoor installations and generally, its size and volume are not limited).

FIGURE 18-5 A metallic IBC storing a flammable dye

FIGURE 18-6 A fiberboard IBC with an internal plastic bag. NFPA 30 prohibits the storage of flammable and combustible liquids in this type of package. *(Courtesy of International Code Consultants, Austin, TX)*

Code Essentials

The term "IBC" can have multiple meanings when dealing with flammable and combustible liquids.
- Normally, IBC refers to the *International Building Code*.
- However, IBC is also used by NFPA 30, U.S. DOT and industry, to refer to an intermediate bulk container.

Portable tanks are a package of more than 60-gallon capacity that are designed for transportation. Portable tanks are equipped with components or features to facilitate material handling. Both DOT and NFPA 30 limit the volume of portable tanks to a volume of 793 gallons (3,000 liters). A portable tank is equipped with a connection on the bottom of the tank to allow for gravity dispensing of the liquid. All portable tanks require some means of pressure relief to prevent the tank from rupturing or exploding if it is involved in a fire. DOT uses the term intermediate bulk container (IBC) to define many portable tanks and permits them to be constructed of metal, plastic, fiber-board or any combination of approved materials. DOT allows the packaging of many flammable and combustible liquids in IBCs constructed of plastic, fiberboard or composite materials. However, NFPA 30 prohibits the packaging of any flammable liquids in portable tanks, drums or IBCs constructed of these materials; only approved metals can be used as the material of construction when storing flammable liquids. NFPA 30 allows the packaging of combustible liquids in rigid plastic and metal IBCs; however, storage of flammable or combustible liquids is prohibited in a composite IBC or a "bag-in-a-box," which essentially is a plastic bag inserted into a fiberboard box (Figures 18-5 and 18-6). **[Ref. 202, 5704.3.1]**

A stationary tank is designed for permanent installation and is not intended for attachment to a transportation vehicle as a part of its normal operation or use. Stationary tanks are designed for above-ground or underground installation and can be installed inside of buildings. Tanks can be field erected or fabricated in a shop and can be constructed with integral secondary containment to contain any leaks from the primary tank. Tanks are designed and installed in accordance with the IFC and NFPA 30, and with few exceptions, the size and volume of a storage tank is not limited (Figure 18-7). Table 18-3 compares the characteristics of the terms container, drum, portable tank and stationary tank. **[Ref. 202, 5704.2.7]**

FIGURE 18-7 A stationary above-ground storage tank

TABLE 18-3 Tank and container terminology

Storage terminology	Volume	Fixed or portable	Typical refilling method
Container	≤ 60 gallons	Portable; nonfixed	Remove empty & insert full
Drum	≤ 119 gallons	Portable; nonfixed	Remove empty & insert full
Portable tank	> 60 gallons and ≤ 793 gallons	Portable; nonfixed	Remove empty & insert full
Stationary tank	Any size	Fixed in position	Add product to tank

DESIGN AND CONSTRUCTION OF STORAGE TANKS

Flammable and combustible liquid storage tanks are designed and constructed in accordance with Underwriters Laboratories (UL) or American Petroleum Institute (API) standards as well as the requirements in NFPA 30 and the IFC. All storage tanks have features that are common among any design:

- The material of construction is chemically compatible with and designed to support the weight of the stored liquid.

- The tank has a normal vent to relieve pressures that are developed when product is introduced or withdrawn.

- Storage tanks operate at a pressure of less than 15 PSIG. A stationary container designed to operate at a pressure above 15 PSIG is a pressure vessel and must be constructed in accordance with the ASME *Boiler and Pressure Vessel Code*.

- A nameplate is attached to the tank that indicates the standard to which it was constructed (Figure 18-8). [Ref. 5704.2.7]

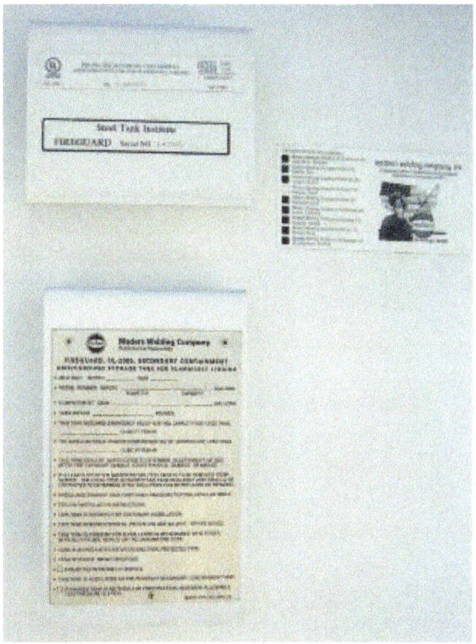

FIGURE 18-8 Nameplate for a protected above-ground storage tank. The nameplate indicates the UL standard that tank was built to and any special installation and operation limitations.

The design requirements vary based on whether the tank is intended for underground or above-ground installation and its operating pressure. Tanks are designed for atmospheric pressure or low-pressure service. Atmospheric tanks store liquid at atmospheric pressure (14.7 PSIA) up to 0.5 PSIG. Low-pressure tanks store liquids at pressures from 0.5 to 15 PSIG. Low-pressure tanks are used predominantly for storing very volatile liquids or liquids that may also be mixed with gases. [Ref. 5704.2.7]

Petroleum storage tanks fall into two categories: shop-fabricated and field-erected. Shop-fabricated storage tanks are fabricated in a manufacturing plant and shipped as a finished assembly to the customer. Field-erected storage tanks are manufactured from plate steel cut and formed into different shapes (Figure 18-9). These

FIGURE 18-9 Construction of a field-erected above-ground storage tank

parts are then shipped to the location where the tank is assembled and tested on site. All of the tanks listed to UL standards are shop-fabricated tanks. Tanks constructed to API standards are either shop-fabricated or field erected. However, the predominance of storage tanks constructed to API standards are field erected.

When compared to shop-fabricated above-ground storage tanks (ASTs), field-erected ASTs are subject to more rigorous inspection requirements. The reason is that neither the IFC nor NFPA 30 limits the volume or diameter of a field-erected AST. Conversely, shop-fabricated ASTs are generally limited to around 50,000–70,000 gallons because the tank has to be assembled and tested in the factory as a condition of its listing. The finished tank is then transported over highways, which limits the size and weight of loads that can be transported. Table 18-4 summarizes the major differences between shop-fabricated and field-erected ASTs.

TABLE 18-4 Design and construction differences between field-erected and shop-fabricated above-ground storage tanks

Variable	Field-erected AST	Shop-fabricated AST
Volume of liquid	Diameter and height of the tank are not limited	Volume is limited to between 50,000 – 70,000 gallons and transportation laws
Design practices	Tank bottom is the foundation and contains the thickest metal	Tank heads are constructed of metal that is thicker than the tank shell
Secondary containment	Normally constructed inside of a containment structure with engineered dike walls and foundation	Constructed in the factory and can be listed as having integral secondary containment
Inspections	Subject to periodic internal and external inspections in accordance with American Petroleum Institute standards	Inspections occur as part of the listing process
Approval	Designed by a registered Professional Engineer and approved by the owner and fire code official	Listed by a nationally recognized testing laboratory and is approved by the fire code official

Storage tanks are constructed to one of the following standards:
- API 12B, Bolted Tanks for Storage of Production Liquids
- API 12D, Field Welded Tanks for Storage of Production Liquids
- API 12F, Shop Welded Tanks for Storage of Production Liquids
- API 620, Design and Construction of Large, Welded, Low-Pressure Storage Tanks
- API 650, Welded Tanks for Oil Storage

- Steel Tank Institute SP01, Standard for Aboveground Tanks
- UL 58, Standard for Steel Underground Storage Tanks for Flammable and Combustible Liquids
- UL 80, Standard for Steel Underground Storage Tanks for Flammable and Combustible Liquids
- UL 142, Standard for Steel Aboveground Tanks for Flammable and Combustible Liquids
- UL 1316, Glass-Fiber-Reinforced Plastic Underground Storage Tanks for Petroleum Products, Alcohols and Alcohol-Gasoline Mixtures
- UL 2085, Standard for Protected Aboveground Tanks for Flammable and Combustible Liquids
- UL 2245, Standard for Below-Grade Vaults for Flammable Liquid Storage Tanks.

Underground storage tanks (USTs) are constructed of carbon steel or fiberglass-reinforced thermosetting plastic and are designed for burial. USTs provide the highest level of fire safety because the tank has no possibility of being exposed to a fire (Figure 18-10). Because a UST is buried, the IFC requires these tanks to be provided with a primary and secondary containment. The primary containment is the tank itself, and the secondary containment can be provided by another tank shell or other approved containment systems. The outer (secondary) containment will capture any leaking product that escapes the primary tank and prevent the product from contaminating soil or a water body. These tanks are equipped with an overfill protection system and electronic systems for leak detection and product reconciliation. Product reconciliation is an important part of an overall spill prevention program as it ensures each gallon of petroleum product is accounted for through receipt, sales or evaporative losses. If the UST is constructed of steel, it is connected to a cathodic protection system to protect it from corrosion.

FIGURE 18-10 Installation of an underground storage tank *(Courtesy of The Steel Tank Institute, Lake Zurich, IL)*

Above-ground storage tanks (ASTs) are constructed of carbon steel or other ferrous materials (Figure 18-11). They can be installed outside or inside of buildings. Because an AST can be easily inspected, it normally does not require cathodic protection. Based on the type of tank and its application, tank connections, known as nozzles, may be above or below the liquid level. Even though ASTs can be visually inspected, almost all installations will require secondary contain-

FIGURE 18-11 An above-ground storage tank at a manufacturing plant

ment. This protection is required to prevent any leaks from forming a large spill of product, which could be the fuel for a pool fire. Except for tanks larger than 12,000 gallons storing Class IIIB combustible liquids, every AST requires an emergency vent, which protects the tank from a pressure explosion if it is involved in or exposed to fire. **[Ref. 5704.2.7.4]**

ASTs and USTs are commonly used for the storage of vehicle fuels including unleaded gasoline, gasoline-ethanol mixtures, diesel fuel and biodiesel. ASTs are commonly specified by many petroleum product suppliers and retailers because they are subject to fewer permit fees that fund government-mandated leaking UST payment programs. The requirements for the type of tank allowed for motor fuel-dispensing facilities are specified in IFC Chapter 23. Since many motor fuel-dispensing facilities are accessible by the general public, the IFC prescribes the use of either ASTs located in below-grade vaults or protected above-ground storage tanks (PAST) when storing Class I liquids above ground and outside of buildings. When a PAST is used, the tank must be installed in accordance with the requirements in Chapter 57. **[Ref. 2306.2.3]**

FIGURE 18-12 A 10,000-gallon protected above-ground storage tank at a fleet vehicle motor fuel-dispensing facility

A PAST is a shop-fabricated tank listed in accordance with UL 2085, Standard for Protected Aboveground Tanks for Flammable and Combustible Liquids. A PAST consists of a primary tank that is constructed to the requirements of UL 142 and is protected from physical damage such as impact by a vehicle (Figure 18-12). PASTs are designed to operate at atmospheric pressure and are constructed with a method of secondary containment that is integral to the storage tank. The tank is constructed with a fire-resistive feature to protect it from a liquid pool fire. The tank may provide all these protection elements as a single unit, may be an assembly of components or may be a combination of both. **[Ref. 5704.2.9.7]**

> ### Code Essentials
>
> Stationary storage tanks are constructed in accordance with recognized national standards approved by the fire code official. Underground storage tanks are shop fabricated. Above-ground storage tanks are field-erected or shop-fabricated. Shop-fabricated tanks are listed by a nationally recognized testing laboratory, while field-erected tanks are approved by the fire code official and the owner of the storage tank. •

TABLE 18-5 Minimum separation requirements for above-ground tanks (IFC Table 2306.2.3)

Tank type	Individual tank capacity (gallons)	Minimum distance from nearest important building on same property (feet)	Minimum distance from nearest fuel dispenser (feet)	Minimum distance from lot line that is or can be built upon, including the opposite side of a public way (feet)	Minimum distance from nearest side of any public way (feet)	Minimum distance between tanks (feet)
Class I protected above-ground tanks	Less than or equal to 6,000	5	25[a,c]	15	5	3
	Greater than 6,000	15	25[a,c]	25	15	3
Tanks in vaults	0 – 20,000	0[b]	0	0[b]	0	Separate compartment required for each tank
Other tanks	All	50	50	100	50	3

For SI: 1 foot = 304.8 mm, 1 gallon = 3.785 L.
a. At fleet vehicle motor fuel-dispensing facilities, a minimum separation distance is not required.
b. Underground tanks shall be located such that they will not be subject to loading from nearby adjacent structures, or they shall be designed to accommodate applied loads from existing or future structures that can be built nearby.
c. For Class IIIB liquids in protected above-ground tanks, a mimimum separation distance is not required.

Installation requirements for ASTs are found in IFC Chapters 23 and 57. IFC Table 2306.2.3 (Table 18-5) sets forth the requirements for locating storage tanks for fuel dispensing based on the type of tank, its volume and its location in relation to certain exposures, including important buildings on the same property, the dispenser, property lines and public ways. **[Ref. 2306.2.3]**

When a PAST is used at a motor fuel-dispensing facility, it must comply with the requirements in Chapters 23 and 57. The PAST also must be equipped with an overfill prevention device and spill containment (Figure 18-13). An overfill prevention device prevents the PAST from being overfilled with flammable or combustible liquids. The device is attached to a liquid-tight tank fill connection. An overfill prevention device is designed with a valve that floats on the liquid inside the tank. As the liquid level is increased during filling, the valve begins to restrict the liquid when the PAST is filled to 90 percent of its rated volume. The valve continues to restrict the flow until the PAST is filled to 95 percent of its rated volume—at this point, the flow of liquid is stopped by the valve. To further limit the potential for a

FIGURE 18-13 A spill container installed on a protected above-ground storage tank. It encloses the tank fill connection, which is equipped with an overfill prevention device.

flammable or combustible liquid release, the fill connection is located inside of a spill container. The spill container is designed to capture any liquid that remains in the hose supplying the fuel from a cargo tanker. A spill container will retain at least 5 gallons of liquid and is equipped with a valve to drain any spilled liquid into the PAST. **[Ref. 5704.2.9.7.5, 5704.2.9.7.7]**

STORAGE TANK OPENINGS

A tank has several openings to accommodate simultaneous tank filling and product withdrawal as well as other components. Components installed on storage tanks must comply with the applicable IFC and NFPA requirements. USTs and ASTs are equipped with at least two openings. One opening is designed to allow the tank to vent vapor to the atmosphere when product is introduced into the storage tank. It allows outside air inside of the tank when product is withdrawn during transfer or dispensing. This opening is called the normal vent. A second opening is provided for introducing or withdrawing product from the tank. **[Ref. 5704.2.7.3, 5704.2.7.5]**

Normal vent

Storage tanks are designed to resist the positive and negative pressures generated when liquid is introduced into or withdrawn from the tank (Figure 18-14). Improper sizing of a tank's normal vent or obstructions in it can generate excessive pressure, which can damage the tank. Normal vents on storage tanks storing Class I, II and IIIA liquids are terminated outside of buildings at least 12 feet above the adjacent ground level. The 12-foot elevation of the vent is necessary to ensure that the surrounding air mixes with the vapor being exhausted so it is diluted below 25 percent of the liquid's LFL prior to the vapor

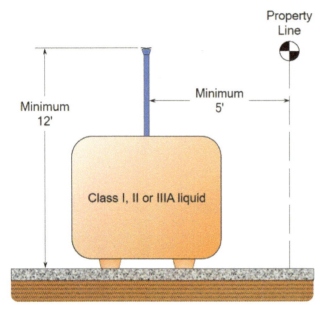

FIGURE 18-14 Normal venting of an above-ground storage tank

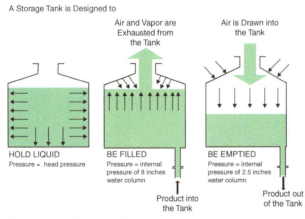

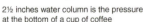
2½ inches water column is the pressure at the bottom of a cup of coffee

8 inches water column is the pressure at the bottom of a glass of beer

FIGURE 18-15 Functions of the normal vent

reaching the ground and drifting away from the vent opening. The vent piping is arranged so vapors are discharged upward or away from property lines. The normal vent is terminated at least 5 feet from building openings or lot (property) lines that can be built upon (Figure 18-15). Vent pipes on tanks storing Class IIIB liquids inside buildings are allowed to terminate inside the buildings provided the vents are normally closed. [Ref. 5704.2.7.3.3]

The normal vent on a tank storing Class I liquids must be equipped with a pressure-vacuum (PV) vent (Figure 18-16). A PV vent is designed to remain in a normally closed position and only operates when the tank is operating under positive pressure or negative pressure (vacuum) conditions. The IFC specifies a PV vent to minimize the loss of vapor when the tank is atmospherically heated. On tanks storing Class IB and IC liquids, an alternative to the PV vent is a flame arrester, which is designed to remain closed except under pressure or vacuum conditions. [Ref. 5704.2.7.3.2]

FIGURE 18-16 The three normal vent pipes on the left are connected to underground storage tanks and fitted with pressure-vacuum vents.

Emergency vent

A third opening is provided for an emergency vent on all ASTs storing Class I, II and III liquids. The emergency vent opening will safely discharge vapors that will be generated if the tank is involved in a pool fire or is subject to an exposure fire. An emergency relief vent is an opening, a construction method or device that relieves internal pressure that can develop inside of a storage tank when it is exposed to fire. An emergency vent is not required for ASTs that contain more than 12,000 gallons of Class IIIB liquids, provided the tank is not located with the containment dike or drainage path of tanks containing Class I or II flammable or combustible liquids. [Ref.5704.2.7.4]

A number of notable incidents have occurred in the United States involving the failure to provide, or the disabling of, emergency vents on ASTs that caused the death of fire fighters. One of the more devastating incidents was a fire at a small petroleum terminal and motor fuel-dispensing station that occurred August 18, 1959, in Kansas City, Kansas. The fire involved a nominal 20,000-gallon horizontal AST. The tank was not listed nor found to be constructed in accordance with any API or UL standards. The tank had a 1½-inch normal vent but was not equipped with an emergency vent. Approximately 90 minutes into the fire-fighting operations, the storage tank catastrophically failed because of the internal pressure and the weakening of one end of the storage tank that was not being cooled by fire streams. When the tank failed, it was propelled through two brick walls and approximately 95 feet into the fire fighters, who were operating large hose streams onto the fire. The fire and tank failure killed five fire fighters and one civilian, and injured over 65 persons. This was the largest number of line-of-duty deaths experienced at that time by the Kansas City (KS) Fire Department. Calculations of the wetted surface area of the tank found that it should have been equipped with a 10-inch emergency vent.

On July 31, 1968, a similar incident occurred outside of Kennadale, Texas. The fire involved a 10,000-gallon homemade AST storing gasoline. The tank was divided into two compartments: 7,000 and 3,000 gallons. The 7,000-gallon compartment was equipped with an overfill opening that acted as an emergency relief vent; however, the size of the opening was undersized by 50 percent. The 3,000-gallon compartment did not have an emergency vent. Approximately one hour into the incident, the storage tank exploded, killing two fire fighters and one bystander, and injuring 57 fire fighters and civilians.

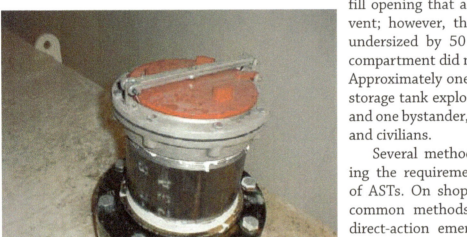

FIGURE 18-17 Direct-action emergency vent. This particular manufacturer uses a spring-loaded cover that is secured closed by a graphite pin. The pin breaks at an opening pressure of approximately 1 PSIG, causing the vent cover to open.

Several methods are available for satisfying the requirements for emergency venting of ASTs. On shop-fabricated ASTs, the most common methods are the installation of a direct-action emergency vent or a long-bolt emergency vent. A direct-action emergency vent is constructed to a minimum diameter and a cover is attached to a metal guide (Figure 18-17). The internal diameter of the direct-action emergency vent governs its pressure relief flow capacity. The weight of the cover governs its opening pressure, which cannot exceed 2.5 PSIG under the tank design requirements. Direct-action emergency vents offer a reliable means of emergency venting for ASTs because the cover is the only moving part.

Another acceptable method of providing an emergency vent is installing a long-bolt vent (Figure 18-18). A long-bolt emergency vent is simply a flange cover mounted onto a flanged manway opening on top of an AST. Every other bolt is removed from the flange cover and the remaining bolts are installed so they have a minimum of 1½ inches of movement. These vents are less reliable because it is possible for the cover to be bolted tight to the flanged manway opening, which disables it. They are

FIGURE 18-18 Long-bolt emergency vent

also subject to warping, especially in hot climates when the metal is cooled by rain.

NFPA 30 requires that all emergency vents be marked to indicate the vent's flow rate. This requirement applies to commercially manufactured and listed emergency vent devices and also to a long-bole vent. The required flow rate to adequately vent a tank under fire exposure, expressed in standard cubic feet/hour (SCFH), is based on the wetted area of the AST. The wetted area is the internal tank area made wet by the stored liquid. Calculation of the wetted area is based on the shape of the storage tank and if it is designed for a horizontal or vertical installation.

The nameplate for UL-listed ASTs is marked to indicate the required emergency vent flow rate. An inspection of any AST should include a comparison of the provided emergency vent flow rate to the required flow rate on the tank nameplate. An emergency vent

FIGURE 18-19 The flow rate of an emergency vent must be marked on the device.

is undersized if the flow rate does not equal or exceed the required rated flow capacity on the storage tank nameplate. NFPA 30 permits adding the rated flow capacity through the normal vent to the rated capacity of the emergency vent to satisfy the total required flow rate indicated on the tank nameplate. In this case, the normal vent must also be marked to indicate its rated flow capacity (Figure 18-19).

Glossary

A

addition – An extension or increase in floor area, number of stories or height of a building or structure.

alteration – Any construction or renovation to an existing structure other than a repair or addition.

appliance – Any apparatus or equipment that utilizes gas as a fuel or raw material to produce light, heat, power, refrigeration or air conditioning.

approved – Acceptable to the fire code official.

automatic sprinkler system – For fire protection purposes, an integrated system of underground and overhead piping designed in accordance with fire protection engineering standards. The system includes a suitable water supply. The portion of the system above the ground is a network of specially sized or hydraulically designed piping installed in a structure or area, generally overhead, and to which automatic sprinklers are connected in a systematic pattern. The system is usually activated by heat from a fire and discharges water over the fire area.

B

boiling point – The temperature at which the vapor pressure of liquid equals the atmospheric pressure of 14.7 pounds per square inch absolute (PSIA) or 760 mm of mercury. Where an accurate boiling point is unavailable for the material in question, or for mixtures that do not have a constant boiling point, for the purposes of this classification, the 20-percent evaporated point of a distillation performed in accordance with ASTM D86 shall be used as the boiling point of the liquid.

branch circuit – That part of an electric circuit extending beyond the last circuit breaker or fuse. The branch circuits start at the breaker box and extend to the electrical devices connected to the service. Branch circuits are the last part of the circuit supplying electrical devices. These circuits are classified in two different ways, according to the type of loads they serve or according to their current-carrying capacity.

building official – The officer or other designated authority charged with the administration and enforcement of the building code, or a duly authorized representative.

C

carbon dioxide enrichment system – A system where carbon dioxide gas is intentionally introduced into an indoor environment, typically for the purpose of stimulating plant growth.

catalyst – A substance that initiates or accelerates a chemical reaction without itself being affected.

change of occupancy – A change in the purpose or level of activity within a building that involves a change in occupancy classification or a change in application of the requirements of the code.

code – A written set of rules, principles or laws.

combustible dust – Finely divided solid material that is 420 microns or less in diameter and which, when dispersed in air in the proper proportions, could be ignited by a flame, spark or other source of ignition. Combustible dust will pass through a U.S. No. 40 standard sieve.

commodity – A combination of products, packing materials and containers.

construction documents – The written, graphic and pictorial documents prepared or assembled for describing the design, location and physical characteristics of the elements of the project necessary for obtaining a permit.

corrosive – A chemical that causes visible destruction of, or irreversible alterations in, living tissue by chemical action at the point of contact. A chemical shall be considered corrosive if, when tested on the skin of albino rabbits by the method described in DOTn 49 CFR 173.137, such chemical destroys or changes irreversibly the structure of the tissue at the point of contact following an exposure period of 4 hours. The term does not refer to action on inanimate surfaces.

D

deflagration – An exothermic reaction, such as the extremely rapid oxidation of a flammable dust or vapor in air, in which the reaction progresses through the unburned material at a rate less than the velocity of sound. A deflagration can have an explosive effect.

detonation – An exothermic reaction characterized by the presence of a shock wave in the material which establishes and maintains the reaction. The reaction zone progresses through the material at a rate greater than the velocity of sound. The principal heating mechanism is one of shock compression. Detonations have an explosive effect.

dewar – A double-walled flask of metal or silvered glass with a vacuum between the walls, used to hold liquids at well below ambient temperature.

dispensing – The pouring or transferring of any material from a container, tank or similar vessel, whereby vapors, dusts, fumes, mists or gases are liberated to the atmosphere.

E

emergency voice/alarm communications – Dedicated manual or automatic facilities for originating and distributing voice instructions, as well as alert and evacuation signals pertaining to a fire emergency, to the occupants of a building.

exit – That portion of a means of egress system between the exit access and the exit discharge or public way. Exit components include exterior exit doors at the level of exit discharge, interior exit enclosures, interior exit ramps, exit passageways, exterior exit stairways, exterior exit ramps and horizontal exits.

exit access – That portion of a means of egress system that leads from any occupied portion of a building or structure to an exit.

exit discharge – The portion of a means of egress system between the termination of an exit and a public way.

F

fire – A chemical reaction that releases heat and light and is accompanied by flame, especially the exothermic oxidation of a combustible substance.

fire area – The aggregate floor area enclosed and bounded by fire walls, fire barriers, exterior walls or horizontal assemblies of a building. Areas of the building not provided with surrounding walls shall be included in the fire area if such areas are included within the horizontal projection of the roof or floor next above.

fire code official – The fire chief or other designated authority charged with the administration and enforcement of the code, or a duly authorized representative.

fire-flow – The flow rate of a water supply, measured at 20 pounds per square inch (psi) residual pressure that is available for fire fighting.

fire protection system – Approved devices, equipment and systems or combinations of systems used to detect a fire, activate an alarm, extinguish or control a fire, control or manage smoke and products of a fire or any combination thereof.

fire resistance – That property of materials or their assemblies that prevents or retards the passage of excessive heat, hot gases or flames under conditions of use.

flashover – An event during a fire's growth where the hot smoke layer inside a room or compartment releases the greatest amount of convective and radiant energy.

H

heat release rate – A measurement of the rate a combustion reaction produces heat and calculated by multiplying the effective heat of combustion of a material by its mass loss rate. It is expressed in British Thermal Units [Btu]/minute or kilowatt (kW).

high-piled combustible storage – Materials in closely packed piles or combustible materials on pallets, in racks or on shelves where the top of storage is greater than 12 feet in height. When required by the fire code official, high-piled combustible storage also includes certain high-hazard commodities, such as rubber tires, Group A plastics, flammable liquids, idle pallets and similar commodities, where the top of storage is greater than 6 feet in height.

hood – An air intake device used to capture by entrapment, impingement, adhesion, or similar means, grease, moisture, heat and similar contaminants before they enter a duct system.

I

inspection – A formal or official examination.

interlock – A method of preventing undesired states in a machine, which in a general sense can include any electrical, electronic or mechanical device or system.

L

laboratory suite – A fire-rated enclosed laboratory area that will provide one or more laboratory spaces, within a Group B educational occupancy, that are permitted to include ancillary uses such as offices, bathrooms and corridors that are contiguous with the laboratory area.

level of exit discharge – The story at the point at which an exit terminates and an exit discharge begins.

liability – The state of being legally obliged and responsible.

lower flammable limit (LFL) – The minimum concentration of vapor in air at which propagation of flame will occur in the presence of an ignition source. The LFL is sometimes referred to as LEL or lower explosive limit.

M

manometer – An instrument for measuring differences of pressure. The weight of a column of liquid enclosed in a tube is balanced by the pressures applied at its opposite ends, and the pressure difference is computed from the hydrostatic equation.

minimum explosive concentration (MEC) – The lowest concentration of dust cloud that will allow combustion. May also be referred to as the lower explosive limit.

mobile fueling – The operation of dispensing liquid fuels from tank vehicles into the fuel tanks of motor vehicles. Mobile fueling may also be known by the terms "mobile fleet fueling," "wet fueling" and "wet hosing."

O

occupant load – The number of persons for which the means of egress of a building or a portion thereof is designed.

on-demand mobile fueling – The operation of delivering a liquid fuel to a vehicle and dispensing the fuel from tank vehicles into the fuel tanks of motor vehicles. The vehicle owner calls to schedule the fuel delivery and specifies the location of the vehicle.

P

performance-based design – An engineering approach to design elements of a building based on agreed-upon performance goals and objectives, engineering analysis and quantitative assessment of alternatives against the design goals and objectives using accepted engineering tools, methodologies and performance criteria.

pressure, residual – The available pressure inside of pipe while the liquid is flowing.

pressure, static – The pressure inside of a pipe or container while the liquid is at rest.

R

retroactive – Affecting things past.

S

spray booth – A mechanically ventilated appliance of varying dimensions and construction provided to enclose or accommodate a spraying operation and to confine and limit the escape of spray vapor and residue and to exhaust it safely.

spray room – A room designed to accommodate spraying operations and constructed in accordance with the *International Building Code*.

spraying space – An area in which dangerous quantities of flammable vapors or combustible residues, dusts or deposits are present due to the operation of spraying processes. The fire code official is authorized to define the limits of the spraying space in any specific case.

standard – Something set up and established by authority as a rule for the measure of quantity, weight, extent, value or quality.

storage – The keeping, retention or leaving of hazardous materials in closed containers, tanks, cylinders or similar vessels; or vessels supplying operations through closed connections to the vessel.

supervisory signal – A signal indicating the need of action in connection with the supervision guard tours, the fire suppression system or equipment, or the maintenance features of related systems.

U

upper flammable limit (UFL) – The maximum concentration of vapor in air at which propagation of flame will occur in the presence of an ignition source.

use-closed system – The use of a solid or liquid hazardous material involving a closed vessel or system that remains closed during normal operations where vapors emitted by the product are not liberated outside of the vessel or system and the product is not exposed to the atmosphere during normal operations; and all uses of compressed gases.

use-open system – The use of a solid or liquid hazardous material involving a vessel or system that is continuously open to the atmosphere during normal operations and where vapors are liberated, or the product is exposed to the atmosphere during normal operations.

W

water-miscible – The property of a substance to mix with water and form a homogeneous solution.

wildland – An area in which development is essentially nonexistent, except for roads, railroads, power lines and similar facilities.

Index

A

Above-ground storage tanks (ASTs), 169, 266-271
Access roads, 59–62
Access to buildings, 62-63
Acetylene, 214
    pressure regulator, 214
Arcolein, 229
Airflow balance testing, 185
Alarms. *See* Fire alarm, detection systems
Alcohol-blended fuels, 173-174
Alternative construction methods, 24-26
American Petroleum Institute (API), 267-269
    See also specific API standards
American Society of Mechanical Engineers, 248-249
    See also specific ASME standards
ANSI Z83.26, 74
API 12B, 268
API 12D, 268
API 12F, 268
API 620, 268
API 650, 268
Appendices, to codes, 10, 19, 21
Appendix A, 39
Appendix B, 21, 65
Appendix D, 59, 60-61
Appendix E, 225
Appendix F, 241
Appendix G, 21
Appendix H, 229

Appliances
    fuels used in, 71-72
    installation requirements for, 71
    portable outdoor, 74-75
    venting for, 73-74
ASME A13.1, 252
ASME A17.1, 77
ASME *Boiler and Pressure Vessel Code*, 267
Assemblies, public, 51
ASTM International, 16
    See also specific ASTM standards
ASTM E84, 96-97
ASTM E1537, 99
ASTM F2200, 61-62
ASTs. *See* Above-ground storage tanks (ASTs)
Authority, of IFC, 22-27
Automated storage, 202-204
Automatic transfer switch, 86-87

B

Batteries, stationary storage systems, 90-91
Beverage dispensing, 259-260
Board of Appeals, 39-40
Breach of duty, 35-36
Buildings
    access to, 62-64
    fuel-fired appliances in, 71
    historic, 15-16
    vacant, 34, 46-47
Burst disc, 250

C

Cannabis oil, 218
Capacitor energy storage systems, 92
Carbon dioxide
    beverage dispensing, for, 259-260
    enrichment, 218-219
Carbon disulfide, 264
Carbon monoxide, 74, 94-95
    alarms, 143-144
Carousel storage, 203-204
cdpACCESS, 4-5
Certificate of Occupancy, 14-15
Change of occupancy, 14, 15
Chemical Abstracts Service (CAS) number, 222, 230
Chopper gun, 182
Codes
    adoption of, 19-22
    amendments to, 19-20, 61
    development, 3-5
    scope, 5-10
    and standards, 16-17
See also individual codes, e.g. *International Fire Code* (IFC)
Combustible dust-producing operations, 207-209
Combustible liquids. *See* Liquids, flammable and combustible
Combustible materials, 43-44
Commodity, 192-195
Compressed Gas Association (CGA)
    CGA Standard C-7, 252

281

CGA Standard 790, 74-75
CGA Standard S-1.1, 249
CGA Standard S-1.2, 249
CGA Standard S-1.3, 249
Compressed gases
  cylinders, containers, tanks for, 248-249
  exhausted enclosures, 254, 256
  and leaks, damage, corrosion, 256-257
  markings for, 251-252
  nesting, 254
  nonliquefied, liquefied, dissolved, 246-247
  pressure relief devices (PRD) for, 249-250
  security, 253
  separation from hazardous conditions, 254-255
  valve protection for, 253-254
Conflicting requirements, 6, 16-17, 146
Construction, and fire safety, 210-212
Construction, retroactive requirements, 12, 13
Construction permits, documents, 29-31
Containers, gas, 248-250
Control area, 239-241
Control-mode sprinklers, 119
Cooking, 79-83
  exhaust hood, 79-81
  fire-extinguishing system, 82-83
  in mobile food preparation vehicles, 49
  oil storage, 81-82
Crowd managers, 51-52
Cryogenic fluids, 247, 251
Cylinders, gas, 248-250
Cylinder exchange program, 257-259

D

Damages, 35-36
Decorative materials, 94-95, 100
Deflagration, 181, 207-209
Deluge automatic sprinkler system, 120, 122, 124
Demolition, and fire safety, 210-212
Department of Fire Prevention, 22
Differential backpressure valve, 175-176
Differential pressure manometer, 186
Direct-action emergency vent, 274
Dispensing, fuel, 165-168
Displays, indoor, 48
Distinct hazard, 11-13, 16
Doors, and egress, 148, 157-160
DOT *Hazardous Materials Emergency Response Guidebook*, 252
Dry-chemical alternative fire-extinguishing systems, 106, 109, 189
Dry-pipe automatic sprinkler system, 120-122
Duct smoke detector, 134-135
Dumpsters, 43-44
Dust-producing operations, 207-209
Duty, 35-36

E

Early Suppression Fast Response (ESFR) sprinklers, 119-120
Egress, means of
  aisles, 158
  doors, 148, 157-160
  elevators as, 54, 76-79, 85, 149
  exit, 146-147
  exit access, 146-147, 153-158
  exit discharge, 146-148
  exit signs, 84, 87, 158-159
  features, characteristics of, 148-149
  illumination, 87, 158-159
  level of exit discharge, 115
  maintenance for, 160
  number of exits, 147, 156-157
  and occupant load, 149-151, 155-156
  separation of exits, 157-158
  travel distance, 148, 153-155
  widths for, 152-153
Elevators, 76-79, 84, 130, 149
  fire service access elevators, 77, 130
  occupant evacuation elevators, 53-54, 130
Emergency breakaway device, 172, 176
Emergency planning, preparedness, 50-53
Emergency power, 83-87
Emergency responder radio coverage, 66-69
Emergency responders
  access, 62-63
  accidents involving, 64, 273-274
  during emergency operations, 9, 26-27
  and hazard identification, 241-242
  radio coverage for, 66-69
  safety of, 9, 27
  using HMMP documents, 230

Emergency voice/alarm communication system, 84, 137, 138-139, 142, 152
Energy storage systems, 90-91
Evacuation plans, 52-53
Evaluation Service Reports (ICC-ES), 24-25
Excess flow control valves, 75, 175-176
Exhaust ducts for commercial cooking, 79-81
    cleaning, 81
    inspection, 80
Exhaust ducts for spray booths, 184-186
Exhausted enclosures, 238, 239, 244, 256
Exit, 146-147
    See also Egress, means of
Exit access, 147, 153-158
Exit discharge, 147-148
Exit signs, 84, 87-88, 158-159

F

Field-erected tank, 267-268, 270
Finishes, flammable. *See* Flammable finishes
Fire alarm, detection systems
    about, 132
    control unit for, 136
    design, installation of, 133-135
    fire safety functions, 136
    initiating device for, 136, 137-138
    monitoring, 113, 135
    occupancies requiring, 139-142
    occupant notification appliances, 138
    smoke detectors, 137, 140, 141-142
    standards for, 133
    testing of, 135

Fire area, 103, 116-117
Fire code official, 22-27
Fire code. *See International Fire Code* (IFC)
Fire department connection (FDC), 130-131
Fire department key box, 62-63, 211
Fire fighters
    access doors for, 197, 199
    hazards to, 47, 48, 63-64
    safety of, 9
    See also Emergency responders
Fire–flow, 64-66
Fire lanes, 59-60, 61
Fire Prevention, Department of, 22
Fire protection systems
    defined, 102
    documents, 106-108
    impairment of, 110-112
    inspection, testing, maintenance of, 108-110
    monitoring, 113
    requirements for, 103-105
    Statements of Compliance for, 108
    See also Sprinkler systems *and the specific types of fire-extinguishing systems*
Fire safety, during construction, demolition, 210-212
Fire safety, evacuation plans, 52-53
Fire service elevators, 77
Fire service features, 59-69
Fire watch, 51-52, 110, 212-213
Flames, open, 45-46
Flame Spread Index (FSI), 96-98
Flammable finishes
    about, 179-182
    powder coating, 180-181, 187, 189
    reinforced plastic, 181-182
    spray finishing, 179-180
    See also Spray booths, spray rooms
Flammable liquids. *See* Liquids, flammable and combustible
Flashover phase, 94-95
Flash point temperature, testing, 262-263
Flue spaces, 195, 200-202
FM 4996, 44, 201
Fuel-fired appliances, 71-75
Fuel oil, 72-73
Fusible plugs, 249-250
    See also Compressed gases; Liquids, flammable and combustible

G

Gas cabinets, 238, 256
Gas detection, 144, 177, 219, 260
Gates, and access, 59, 61-62
Generators, for emergency power, 83-84, 87

H

Hazardous Material Expert Assistant (HMEX), 225, 243-244
Hazardous materials
    classification of, 224-229
    control areas for, 239-241
    identification signs for, 241-242
    MAQ for, 233-238
    reporting of, 229-231
    separating incompatible materials, 243-245
    storage of, 223, 231

unauthorized discharge of, 231
use of, 223, 231-232
Hazardous Materials Inventory Statement (HMIS), 229-230
Hazardous Materials Management Plan (HMMP), 229-230
Health hazards, 224, 227-228
Heat release rate
    of high–piled combustible storage, 192, 194
    of interior finish, 97
    of pallets, 44, 201
    of plastics, 194
    of upholstered furniture, mattresses, 99-100
High-hazard commodity, 191-194
High-piled combustible storage
    about, 191
    aisles for, 204-205
    automated storage, 203-204
    commodity classification, 192-195
    designating areas of, 195-196
    fire department access doors, 199
    flue spaces, 195, 200-202
    plastics, 194-195
    sprinkler protection, 196-198
    storing, 200
Higher education
    laboratories, 215-218
    laboratory suites in, 215-216
Hoods, commercial kitchen, 79-81
Hot work, 212-215
    permit program, 211, 213
Hydrants, testing, 64-65
Hydrogen fuel gas rooms, 177
Hydrostatic pressure relief valve, 176

I

IBC. *See* International Building Code (IBC)
ICC. *See* International Code Council (ICC)
ICC Evaluation Service, 24-25
IFC. *See* International Fire Code (IFC)
IFGC. *See* International Fuel Gas Code (IFGC)
Ignition sources, 45
Illumination, at exits, 158-159
IMC. *See* International Mechanical Code (IMC)
Impairment, of fire protection system, 110-112
Incipient fire phase, 94-95
Incompatible materials, 243-245
Indoor displays, 48
Inspections
    about, 33
    and liability, 35-37
    right of entry for, 33-34
    and stop work orders, 38-39
    and testing, equipment, 37
    of unsafe buildings, 37-38
    warrant, 34-35
Interior finish, 96-99
Interlocks, for spray booths, 188
Intermediate bulk container, 265-266
International Building Code (IBC)
    and certificates of occupancy, 14-15
    correlation with IFC, 3, 10
    scope of, 6-7
International Code Council (ICC), 3-5
International Fire Code (IFC)
    adoption of, 19-21
    amending, 19-20, 61
    appendices to, 10, 19, 21
    applicability of, 11-12
    authority in, 22-23, 26-27
    code development, 3-5
    conflicts with referenced standards, 6, 17
    correlation to local, state laws, 21-22
    correlation with IBC, 3, 10
    organization of, 10
    scope, 10-11
International Fuel Gas Code (IFGC), 6, 9, 71, 210
International Mechanical Code (IMC), 6, 8
    on cooking exhaust hoods, 79, 83
    correlation with IFC, 6
    on duct smoke detectors, 134
    and flammable finishes, 182, 185-186
    and fuel-fired appliances, 71, 73
    and hazardous material, 224, 255
    and mechanical refrigeration, 75
    scope, 8
    on temporary heating, 210
International Property Maintenance Code (IPMC), 9
International Residential Code for One- and Two-Family Dwellings (IRC), 3, 7, 118

International Wildland-Urban Interface Code (IWUIC), 8
IPMC. See *International Property Maintenance Code* (IPMC)
IWUIC. See *International Wildland-Urban Interface Code* (IWUIC)

K

Key box, fire department, 62-63, 211

L

Laboratory suite, 215-216
Landscaped roofs, 48-49
Legal aspects, of code enforcement, 19-40
Legal nonconforming, 13
Level of exit discharge, 115
Liability, in inspections, 35-37
Limited spraying spaces, 182-183
Liquefied petroleum gas (LP-gas), 214
  cylinder exchange program, 253, 257-259
  dispensing, 174-176
  excess flow control valve, 75, 175-176
  overfill prevention device (OPD), 257
Liquids, flammable and combustible
  accidents involving, 273-274
  classification of, 262-264
  containers, tanks for, 264-267
  storage tank openings for, 272-273
  tank design, construction, 267-271
Long-bolt emergency vent, 274
Lower flammable limit (LFL), 263-264

Luminaires, for spray booths, 187

M

Marijuana processing, 218-219
Material Safety Data Sheets (MSDS), 225
  See also Safety Data Sheets (SDS)
Mattresses, 99-100
Maximum Allowable Quantity (MAQ)
  defined, 223
  increases in, 238
  per control area, 233-238
  and sprinkler systems, 237-238
  tables for, 234-236
Means of egress. *See* Egress, means of
Methylacetylene-propadiene (MAPP) gas, 214
Methyl ethyl ketone peroxide, 181
Mobile food preparation vehicles, 49
Mobile fueling, *See* On-demand mobile fueling
Model codes, 16, 19
Monitoring fire protection systems, 113
Motor fuel-dispensing facilities
  and alcohol-blended fuels, 173-174
  attended fuel-dispensing, 165-167
  dispensers, 167
  dispensing operations, devices, 165-167
  emergency shutoff switch, 167
  flammable, combustible fuel dispensing, 168-174
  hydrogen fuel dispensing, 176-177

liquefied petroleum gas (LP-gas) dispensing, 174-176
nozzles, fuel delivery, 172, 175-176
requirements by fuel type, 165
storage tanks for, 168-170
unattended fuel-dispensing, 165-167

N

National Fire Protection Association (NFPA), 3
  See also specific NFPA codes
NEC®, See NFPA 70
NFPA 4, 110
NFPA 10, 168
NFAP 12, 189
NFPA 13, 60, 66, 73, 117, 118, 119, 120, 122, 129, 130, 141, 142, 152, 157, 188, 191, 193-194, 196, 198, 201-204, 238
NFPA 13D, 60, 117, 118, 129
NFPA 13R, 17, 60, 66, 117, 118, 129, 130, 142, 152, 157
NFPA 14, 212
NFPA 16, 188
NFPA 17, 189
NFPA 24, 65
NFPA 25, 66, 109
NFPA 30, 16, 72, 81, 165, 261, 263, 265-268, 274-275
NFPA 30A, 165, 261
NFPA 30B, 193
NFPA 31, 72
NFPA 33, 184, 188-189
NFPA 52, 165
NFPA 55, 165, 219
NFPA 57, 165
NFPA 58, 74, 165, 174-176, 257

NFPA 59A, 165
NFPA 70 *National Electrical Code*® (NEC®), 83, 133, 183
NFPA 72, 133, 135-137, 140, 198
NFPA 110, 83, 87
NFPA 111, 83, 87
NFPA 286, 97
NFAP 385, 178
NFPA 652, 209
NFPA 654, 208
NFPA 664, 208
NFPA 704, 241-242, 251
NFPA 750, 189
NFPA 853, 89
NFPA 914, 16
NFPA 1142, 65
NFPA 2001, 189

O

Occupancy classification, and building codes, 7, 14
Occupant evacuation elevators, 53-54, 130
Occupant load, 14, 147-151
Occupational Safety and Health Administration (OSHA), 219, 226, 243
Odorization, of gas, 71
Office of the Fire Marshal, 22
On-demand mobile fueling, 178
Open flames, 45-46
Ordinance, adopting IFC, 19-20
Organic peroxide, 86, 181, 227, 238, 240, 243
Oxygen-acetylene torch, 214
Overfill prevention device (OPD)
 LP-gas, 257
 PAST, 271

P

Palletized storage, 197, 200

Pallets, 200-201
 idle, 191, 194
 outdoor storage, of, 44
 plastic, 44, 201
Paracelsus, 227
PAST. See Protected above-ground storage tank (PAST)
Permits, 12, 15, 27-33
 application for, 29-30
 construction documents, for, 29
 defined, 27
 example of, 32
 fees for, 19, 33
 operational and construction, 27-28
Physical hazards, 226-227
Piping, for fuel dispensing, 170-172
Plant processing, 218-219
Plastics,
 classification of, 194-195
 foam plastic trim, 99
Point of transfer, for fuel, 174-175
Portable unvented heater, 73-75
Powder coating, 180-181, 187, 189
Pre-action automatic sprinkler system, 120, 122-123
Pressure relief device (PRD), 249-250, 266
 mechanical refrigeration, for, 76
 liquefied petroleum gas (LP-gas), for, 174, 176
Pressure-vacuum (PV) vent, 273
Protected above-ground storage tank (PAST), 72-73, 169, 269-271
Proximate cause, 35-36
Public assemblies, 51

Pumps, for fuel dispensing, 170-171, 174-175

R

Rack storage, 200-202
Radiation heat transfer, 204
Radios, public safety. *See* Emergency Responder Radio Coverage
Referenced standards, 16-17
Reinforced plastic, 181-182
Reportable quantity (RQ), 252
Right of entry, for inspectors, 33-34
Rooftop gardens, 48-49
Room-corner fire test, 97

S

Safe dispersal area, 148
Safety Data Sheets (SDS), 56, 225, 237, 243
Scope, of codes, 5-11
See v. The City of Seattle, 34-35
Shafts, identification of, 48, 64
Shop-fabricated tank, 73, 267-268, 270, 274
Smoke detectors, 77-78, 134-135
 See also Fire alarm, detection systems
Smoke-Developed Index (SDI), 96
Solar photovoltaic systems, 88-89
Solid-pile storage, 200
Spray booths, spray rooms
 construction of, 182-184
 fire protection for, 188-189
 illumination for, 187
 interlocks for, 188
 ventilation in, 185-186
Spray finishing, 180-181

Sprinkler systems
    about, 120-122
    control-mode designs, 119
    deluge automatic, 120, 122, 124
    design, installation of, 117-120
    dry-pipe automatic, 120-122
    early suppression fast response (ESFR) sprinklers, 119-120
    exempt locations, 129-130
    and Fire Area, 116-117
    fire department connection (FDC) for, 130-131
    and hazardous materials, 237-238
    and high-piled combustible storage, 196-198
    and interior finish materials, 98
    and level of exit discharge, 115
    monitoring, 113
    occupancies requiring, 122-128
    pre-action automatic, 120, 122-123
    for spray booths, 188-189
    suppression-mode, 119-120
    system designs, 117-120
    throughout, 129-130
    wet-pipe automatic, 120-121
Standards, technical, 16-17
Standby power systems, 83-87
Statement of Compliance, 107-108
Stationary storage battery systems, 90-91
Steel Tank Institute SP01, 269
Steiner Tunnel Test, 96
Stop work order, 38-39
Storage
    of combustible materials, 43
    of compressed gases, 248-249, 253-256
    of cooking oil, 81-82
    of flammable, combustible liquids, 267-271
    of fuel oil indoors, 72-73
    of hazardous material, 231-232
    in higher education laboratories, 215-217
    of liquefied petroleum gas (LP-gas), 74
    of pallets, outdoor, 44
    See also High-piled combustible storage
Sulfuric acid, 91, 226-227
Supervising station, 113, 135-136
Supervisory signals, 133, 135
Suppression-mode sprinklers, 119-120

T

Tanks,
    for compressed gases, 248-249
    for flammable, combustible liquids, 267-271
    vents, normal, 267, 272-273
    vents, emergency, 273-275
Testing
    of egress illumination, 88
    of emergency and standby power, 83, 87
    of fire alarm, detection system, 135
    of fire protection systems, 108-110
    of flash points, 263
    of interior finishes, 96-97
    of radio systems, 67-68
    requirement for, 37, 109
    responsibility for, 37
    of upholstered furniture, mattresses, 99-100
Torch-applied roofing systems, 215
Toxicity, of health hazards, 228
Training,
    of crowd managers, 52
    in emergency evacuation, 54-55
    of employees, 55-56
Travel distance, for egress, 148, 153-155

U

UL 58, 269
UL 80, 72, 73, 82, 142, 269
UL 142, 72, 73, 82, 170, 269, 270
UL 710B, 80
UL 723, 96
UL 924, 159
UL 1316, 269
UL 2034, 143
UL 2075, 143
UL 2085, 72, 73, 169, 269, 270
UL 2152, 82
UL 2196, 85
UL 2227, 258
UL 2245, 170, 269
UL 2335, 44, 201
Underground storage tanks (USTs), 168, 173, 269
Underwriters Laboratories (UL), 3, 96

See also specific UL standards
Unsaturated polyester resin (UPR), 181
Upholstered furniture, mattresses, 99-100
Upper flammable limit (UFL), 173, 176, 263
U.S. Department of Transportation (DOT), 21, 71, 224, 226, 228, 248, 251-252, 264
Use, and hazardous materials, 224, 231-232

V

Vacant premises, 34, 46-47
Vapor recovery, 170
Vapor-processing systems, 167, 172
Vaulted tanks, 168-170
Vents, in storage tanks, 267, 272-275
Violations, to IFC, 11-12, 27

W

Warning placard, 47
Waste oil heaters, 71-72
Water supplies, 64-66
Welding, hot work, 212-215
Wet-pipe automatic sprinkler system, 120-121
Wired communication system, 68-69
Wiring, for emergency power, 85

Discover the ICC Assessment Center, Featuring PRONTO™

All certification and testing activities are now available in the ICC Assessment Center, formerly ICC Certification & Testing.

Take the Test at Your Location

Skip the trip to the testing center for your next ICC Certification exam. Instead, take advantage of ICC PRONTO, an industry leading, secure online exam delivery service. The only proctored remote online testing option available for building professional certifications, PRONTO allows you to take ICC Certification exams at your convenience in the privacy of your own home, office or other secure location. Plus, you won't have to wait days or weeks for exam results, you'll know your pass/fail status immediately upon completion.

 With PRONTO, ICC's Proctored Remote Online Testing Option, take your ICC Certification exam from any location with high-speed internet access.

 With online proctoring and exam security features you can be confident in the integrity of the testing process and exam results.

 Plan your exam for the day and time most convenient for you. PRONTO is available 24/7.

 Eliminate the waiting period and know your results immediately upon exam completion.

 ICC is the first model code organization to offer secured online proctored exams—part of our commitment to offering the latest technology-based solutions to help building and code professionals succeed and advance.

Discover the new ICC Assessment Center, ICC PRONTO and the wealth of certification opportunities available to advance your career: www.iccsafe.org/MeetPRONTO

People Helping People Build a Safer World®

Valuable Guides to Changes in the 2018 I-Codes®

FULL COLOR! HUNDREDS OF PHOTOS AND ILLUSTRATIONS!

SIGNIFICANT CHANGES TO THE 2018 INTERNATIONAL CODES®

Practical resources that offer a comprehensive analysis of the critical changes made between the 2015 and 2018 editions of the codes. Authored by ICC code experts, these useful tools are "must-have" guides to the many important changes in the 2018 International Codes.

Key changes are identified then followed by in-depth, expert discussion of how the change affects real world application. A full-color photo, table or illustration is included for each change to further clarify application.

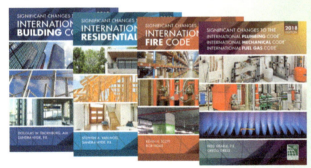

SIGNIFICANT CHANGES TO THE IBC, 2018 EDITION
#7024S18

SIGNIFICANT CHANGES TO THE IRC, 2018 EDITION
#7101S18

SIGNIFICANT CHANGES TO THE IFC, 2018 EDITION
#7404S18

SIGNIFICANT CHANGES TO THE IPC/IMC/IFGC, 2018 EDITION
#7202S18

ORDER YOUR HELPFUL GUIDES TODAY!
1-800-786-4452 | www.iccsafe.org/books

HIRE ICC TO TEACH
Want your group to learn the Significant Changes to the I-Codes from an ICC expert instructor? Schedule a seminar today!

email: **ICCTraining@iccsafe.org** | phone: **1-888-422-7233 ext. 33818**

Accelerate your Building, Plumbing and Energy Products' Speed to Market with ICC-ES!

Your One-Stop Testing, Listing and Product Evaluation Service

- ICC-ES provides a one-stop shop for the evaluation, listing and now testing of innovative building products through our newly formed cooperation with Innovation Research Labs, a highly respected ISO 17025 accredited testing lab with over 50 years of experience.

- ICC-ES Evaluation Reports are the most widely accepted and trusted technical reports for code compliance. When you specify products or materials with an ICC-ES report, you avoid delays on projects and improve your bottom line.

- ICC-ES is a subsidiary of ICC, the publisher of the codes used throughout the U.S. and many global markets, so you can be confident in their code expertise.

- ICC-ES provides you with a free online directory of code compliant products at: **www.icc-es.org/Evaluation_Reports** and CEU courses that help you design with confidence.

WE CERTIFY & TEST:
- Air, Water & Vapor Barriers
- Cladding
- Doors
- Fasteners
- Roofing Materials & Accessories
- Wall Coverings/Systems
- Windows
- Floor/Deck Systems
- Flooring Materials
- Manufacturers Wood
- Plastic Lumber
- Plumbing Products

Look for the ICC-ES Marks of Conformity

WWW.ICC-ES.ORG | 800-423-6587 X3877

Have you seen ICC's Digital Library lately?

codes.iccsafe.org offers convenience, choice, and comprehensive digital options

Get FREE access 24/7

publicACCESS

Enjoy **FREE** access to the complete text of critical construction safety provisions including:

- International Codes®
- ICC Standards and Guidelines
- State and City Codes

premiumACCESS™

In addition to viewing the complete text online, **premiumACCESS** features make it easier than ever to save time and collaborate with colleagues.

Powerful features that work for you:

FREE Demo available!

- Advanced Search crosses your entire set of purchased products.
- Concurrent user functionality lets colleagues share access.
- Internal linking navigates between purchased books in your library.
- Print controls create a PDF of any section.
- Bookmarks can be added to any section or subsection.
- Highlighting with Annotation option keeps you organized.
- Color coding identifies changes.
- Tags can be added and filtered.

1-year and 3-year *premiumACCESS* subscriptions are available now for:
International Codes® | State Codes | Standards | Commentaries

Let codes.iccsafe.org start working for you today!